MAYNARD

Secrets of a Bacon Curer

MAYNARD

Secrets of a Bacon Curer

Maynard Davies

WITH ANN PURCHASE

MERLIN UNWIN BOOKS

First published in Great Britain by Merlin Unwin Books, 2007

Copyright © Maynard Davies, 2007

MERLIN UNWIN BOOKS
Palmers House
7 Corve Street, Ludlow
Shropshire SY8 1DB, U.K.
Tel 01584 877456
Fax 01584 877457
email: books@merlinunwin.co.uk
website: www.merlinunwin.co.uk

British Library Cataloguing in Publication Data:
A catalogue record for this book is available from
the British Library

ISBN 978-1-873674-93-2

Designed and typeset by Merlin Unwin Books, Ludlow, UK.
Printed in Great Britain by Cromwell Press

To my dear wife Ann – thank you

CONTENTS

v

ACKNOWLEDGEMENTS

I am indebted to my wife Ann, for all her help, support and hard work in making this book possible. I also wish to thank all my family and friends who have been supportive at all times. A very special thank you to Karen and Merlin and all the staff at Merlin Unwin Books for having faith, patience and the vision to publish another book by me!

PUBLISHERS' NOTE

The publishers wish to extend particular thanks to Maynard's wife Ann who transcribed his story from the tape recordings he made, because Maynard is dyslexic.

The recipes in this book may not conform in every small detail to contemporary regulations which the modern curer should research themselves.

CHAPTER 1
Goodbye Daisy Bank

We drove up the hill for the last time and bid one final farewell to Daisy Bank, our home for the last 15 years. We had had a lot of happiness there; we had more happiness than money but we thought happiness was a better currency. I gazed round at the hills and down the long track we had just driven and I remembered the first time we had driven down towards our new home with four small children, a very small amount of money and a lot of hope in our hearts. I had left behind a secure job into which I had been apprenticed after leaving school.

By a twist of fate I had become a Master Curer of traditional bacon and hams, because when I left school I could neither read nor write and if it had not been for the chance I was given by an old Curer by the name of Thea, I do not know where I might have ended up. When I had first left school I had tried for other jobs and apprenticeships but as soon as I told them I could not read or write I was shown the door. In those days dyslexia was not acknowledged – I had never even heard of it – and at school I was classed as disruptive and left to my own devices.

One day I turned up at Thea's factory and asked for a job; he took one look at me and told me to clean his office windows. He had handed me some sheets of newspaper and told me to get on with it. He watched my every move while I cleaned the windows

1

inside and out. When I had finished he had growled at me, 'Turn up in the morning at six o'clock, and if you can't be on time, don't bother coming. You will need two pairs of clogs from the clog maker.' I had turned up well before six o'clock the following morning and that was the start of my seven-year apprenticeship into the traditional art of bacon curing and good food.

After many years of learning, at home and abroad, I had met my wife Trisha, a nurse, and we had four lovely daughters. While they were young we decided to buy a farm and set up in business on our own. We had many years of hard work in the Peak District, a very hard part of the country with its very hard winters and as the children grew up and started to leave home, we had decided it was time to move to warmer climes. So we put Daisy Bank on the market and it was duly sold to a very nice couple.

Now we were leaving to begin a new life with two daughters, as one had married and was staying in the Peak District, and the other daughter had gone to train as a nurse. As I looked over the hills I remembered all the happy times and the sad ones too at Daisy Bank but I knew in my heart that now was the right time to move on and seek pastures new. I wondered what life would throw at us next.

We had left the little farm in a better state than we had found it and it was a piece of my life that I will always be grateful for. It gave me a lot of confidence and, for a child born in an industrial town, another look at how life could be.

But now we all got back into the Landrover and drove along the winding country lanes and down from fifteen hundred feet high to sea level. As we were going along the road I thought to myself 'Fortune favours the brave.' It was a new beginning and whatever came along I knew we would tackle with spirit and courage like everything else we had done.

We had already rented a bungalow in the suburbs of a large city and as we drove into the area, I looked at the bungalows, about fifteen of them all clustered together, and the man was a genius who had put them there as you could not put anything between them, they were so close together and I knew from that moment that it would only be a short stay! We moved in temporarily. The neighbourhood ritual on a Sunday was that everyone's doors opened at the same time and all the men came out with their buckets and leathers and set to and cleaned their cars. At twelve o'clock they all went in and ten minutes later they all appeared again and proceeded to the pub. I knew this wasn't for me and yearned for the hills and the animals and my solitude.

My wife Trisha was still nursing at the hospital and because she worked Monday to Friday, at the week-ends we looked for other properties. One Saturday morning we went to a small market town in north Shropshire. The Estate Agent's office was quite untidy, paper everywhere, and a man behind a desk looked up, never said anything, then looked down and carried on working.

I said, 'Excuse me, Sir, we are looking for a property in this area, a small farm on a busy main road.'

He stopped his work this time and said, 'I think we might have one.'

We followed his directions to the property, which stood back about two hundred yards from the main road and it was a little bit unkempt; you could see the grass was overgrown and the hedges needed trimming, but she looked a lovely old house: all she needed was a bit of love; nobody had loved her for a long time.

I thought, 'This could be it,' and I pulled up in front of the house, got out and looked around. You just know, sometimes, when things are right, and I had the feeling that this was right. I could see she was very neglected but I knew we could love her.

We knocked at the door and a man answered it. I told him we had come to view the property and he said, 'Yes I know, the Estate Agent rang me. Come in.' I shook hands with the man but it was an uncomfortable feeling I got straightaway and I thought to myself: 'When you shake hands with this man, you have to count your fingers!' but I was not going to be put off.

We walked through a passageway into a lovely room with a

3

roaring log fire in a lovely old Shropshire grate; in fact it was such a nice scene you could have put it on a Christmas card. We went into the kitchen, a big kitchen but it did need some love. We walked round the house and I thought he would have been better with a mop and bucket, keeping the place clean, but I did not make any comment. As we walked round the whole house I thought it could be made into a home, so I asked the man his name and he told me it was Williams. I asked him if we could go round the out-buildings and he agreed, and after that we walked round the fields.

Trish and I met him back at the house, and we went into the kitchen where we met his wife. I said to the owner that, having discussed it with Trisha, we would like to make him an offer for the property but we could not offer the asking price so I suggested we called the agent down to discuss the deal. The agent came and we agreed on a price and date we could move in.

The wheels were in motion and we had bought ourselves a farm. I thought we could make a go of it although it would be hard work as it was in a very poor state and not been looked after very well, but when it was done I knew it would be a very nice home.

The day came when we moved in and left the bungalow. Trisha asked me to go down to the local shop and settle the newspaper bill, so I went down to there with the dogs in the Landrover. The newsagent was a pleasant man and when I told him we were moving from the area. He said he thought it was a pity because where we lived was nice.

I told him I thought it was very over-crowded and said I was glad we were moving. I explained we were moving to Shropshire and he wished us all much happiness and I thanked him. We shook hands and I felt that marked the start of our new beginning. The Landrover was packed with all the last-minute bits and pieces, the dogs were in the back and we set off to Shropshire.

We arrived at the farm ahead of the furniture vans and it all looked a bit desolate with nobody about but we had made our bed and now we had to lie in it. We went into the house and I was amazed to see that everything that could be taken had been: all the light fittings, the stove, and a lot of the stuff which had not been worth taking – but still he had taken it all.

That was a disappointment. The carpets we had bought with the house had been exchanged for second-hand cheap ones from an auction house as they still had the auction numbers on them. I wondered what else I would find. I walked round all the out-buildings. There had been a fully-equipped dairy and milking parlour but that had gone! He had taken doors off different places in the dairy, taken the milk tank out, demolishing the wall in the process as he could not get it out of the door. He had taken the farm gates, the oil tank: everything that was moveable had gone. I thought to myself, 'Well, I hope fate pays him out in hard currency,' but I wasn't going to be beaten by this.

The furniture van arrived. Trisha told the men where she wanted all the furniture putting and so the unloading began. It was eight o'clock at night before it was all done: we had only put one bed up, so we more-or-less camped out that first night. The next morning, the first priority was to go out and buy a stove as we now had nothing to cook on. Trisha, myself and the children all piled into the Landrover and headed off for the nearest market town.

Trisha went to the supermarket and I went to the electrical shop, explaining to the owner that I wanted a cooker. He showed me one and we agreed a price.

I asked, 'Any chance of you fitting it today? We have just moved into a property and we're a bit desperate as all the cooking equipment has been removed.'

He said, 'Well now, it is already half past ten; we finish at twelve o'clock so it won't be today.'

I said, 'I would consider it a great kindness if you could help us out today as we have no way of cooking anything.'

He turned to me and said, 'Everyone has their problems.'

I thought to myself 'This is hard country we've moved to.' 'If you can't fit it for me today, then would you fit it for me on

Monday and meanwhile sell me a kettle?' I asked – which he did and insisted I paid him there and then. We went back to the farm and made the best of it.

On Monday morning he arrived as arranged. We unloaded the stove, but before he even took off the packing case off he said he would need twenty pounds to connect it to the mains as we had only done the deal on the stove, not the fitting.

'Fair enough,' I said, 'I'll give you the twenty pounds.'

I was in no position to argue as I had to get some cooking equipment for the children and Trisha to have a meal. We paid him and within a few minutes he had connected the cooker to the mains and he asked if he could use the telephone.

I said to him, 'No you can't use the phone: you have given me no courtesy or kindness, you left us for the weekend with no method of cooking a meal. There is a phone down the road – use that!'

I thought that was the right thing to do – fight fire with fire!

We had been in the house about a week and we were not sleeping very well. I used to wake up every night, agitated and perspiring. Trisha and the children couldn't sleep either.

I said to Trisha, 'There's something wrong with this house: we've got a presence.'

She told me not to be silly. Later that morning, Trisha decided she would go and do some shopping for bits and pieces, taking the children with her. As it was a cold day, I decided I would light a fire. I fetched the morning sticks and a full basket of logs and I was just about ready to light the fire when I suddenly went very cold and I could not understand why. I looked behind me and there was a man in a long blue coat with bright buttons on it and he had long golden hair. I blinked and he was gone.

I have never been so frightened in all my life! It was all over in a second, but it really unnerved me; I started to shake. I had never experienced anything like that before and it shook me to the core. When Trisha came back with the children, I told her I had lit the fire and the house was beginning to warm up, but I did not mention what I had seen as I did not want to frighten her and the children.

So I just said to her, 'I've been thinking: we should have the local vicar to bless the house.' Trisha asked me if I thought that was necessary but I just told her I thought it was the right thing to do.

I rang the local vicar, a nice man named Reverend Green. I introduced myself, told him where we lived and said I thought we had a presence. He asked me if I was sure and I told him I was, explaining about the apparition I had seen. I told him I wasn't religious and I hoped that would not put him off.

He said, 'Not at all.'

He said he would come up after the Sunday morning service, and I told him I would appreciate that. He had a nice way with him on the phone. We waited for him on the Sunday morning after the service, Trisha, me and the children. He arrived about half past twelve in a battered old car and I was so glad to see him. He came in, we shook hands and I introduced him to Trisha and the children. He said he would like us all to accompany him in every room where we would all say a prayer together: this was a new experience for me as I am not religious, but I thought on this occasion I might be converted.

We went up the stairs and started in the back bedroom. He put his vestments on and made the sign of the cross and sprinkled some holy water and then we said a prayer. That was the first room done. We walked down the passageway to the next room and repeated the process again, and we did this in all the rooms upstairs.

We then made our way downstairs. As he walked into the dining room, he put up his hand and said, 'I feel there is a presence here.' We all stood in a circle and said a prayer and Reverend Green sprinkled some holy water and made the sign of the cross and that completed the ceremony.

I said to him, 'You don't know how grateful I am, Reverend

Green,' but he just replied, 'It's all part of the job, from the cradle to the grave,' and we both laughed. He had a sense of humour and I find that in life you take to certain people and I took to him. I thought he was a nice man. I asked him how much I owed him and he said, 'Nothing.'

I said, 'Reverend Green, I am a Curer by trade but I haven't started producing anything yet as we have only just moved in. Would you do me the kindness of accepting a ham for Christmas?'

He accepted with thanks. When I started producing and Christmas came I made sure he had the nicest ham.

We had been at the property for about five days and things seemed to be going along smoothly. We were settling down and tidying the place up and it did need it. As it became more comfortable, we got more used to it.

The next morning we were still in bed at about six o'clock when we heard a very loud knock on the door. I looked at Trisha and said, 'That's a very early call, I'll go down and see.' I opened the door and there stood two men with another two men beyond them sitting in a large furniture van.

I said, 'Can I help you?'

One looked at me and said, 'Mr Williams, we have a writ here to take possession of your property.'

I said, 'You've missed Mr Williams by five days.'

The man looked sceptical and said, 'Can you prove who you are?'

I replied, 'At the minute my wife is upstairs. She will prove who I am but I don't suppose you will take that as proof. There's my Mother and my birth certificate but I am sure you will believe none of those either, but if you will give me time I will get in touch with my solicitor and he will come and vouch for me.'

He asked if he could come in and I replied, 'You can come in on the understanding that you will touch nothing and take nothing, as nothing belongs to Mr Williams here. He left five days ago taking his stuff with him - and some of my things too.'

The bailiff laughed. He came in: it was still only about half past six in the morning so I offered him a cup of tea which he gratefully accepted. We waited until nine o'clock and rang my solicitor, who said, 'This is a fine kettle of fish: I think I had better come down.'

So he duly arrived bringing the deeds and the land registration and proof of my identity, and at last the man was satisfied the bird had flown. He asked me if Mr Williams had left anything and I told him there were fifty ewes in the bottom meadow which Mr Williams had left and never returned for.

The bailiff decided to confiscate them, and the next morning a cattle wagon arrived and the ewes were taken off to the auction.

I thought to myself: 'Maynard, welcome to Shropshire!'

CHAPTER 2
Muck and Grand Openings

Now we had to start earning a living curing bacon, so we decided to start on the buildings. Basically, the old barn was seventy feet by forty-five feet and I planned out where I could put all the machinery, where I could site the sausage machine, the curing fridge, the holding fridge, the sink and washing up area, the preparation and slicing area. When I had it all planned out in my mind, one evening I sketched it all out on a Kellogg's cornflake box.

The building was in a dreadful state. The previous owner had left dead chickens lying around, the floor was about eighteen inches thick in muck and I knew this was the first thing to tackle, so I set to with a barrow and I wheeled every bit of muck into the field and spread it. It took me about a fortnight to clean out all the muck and refuse.

After this, the building needed a really good clean, so I looked in the phone book for anyone with a high pressure hose, and found a man with one at £35 to hire for the day, so off I went to hire myself a high pressure hose.

The owner of the tool-hire company said, 'You're a stranger in the area, aren't you?'

I said I was and when I gave him my address, he seemed to freeze up, went into the office, came out and told me I would have to come back with a deposit of £100 on the machine. I asked him why? He said it was company policy with new people moving into the area. Back at the farm I told Trish something was radically wrong because as soon as I mentioned the name of the farm, the tool-company wanted a deposit of £100. I went back and gave him the £35 hire fee and a cheque for the £100.

I thought to myself: this is going to be a hard climb, a very hard climb; every hurdle seems to get higher, but I wasn't going to let this put me off. So I gave everywhere a good clean and powered-hosed it all through. It took me three days to clean it properly, and then I disinfected it all the way through.

Next on my list was to clean all the drains to make sure they were running properly. In fact they where all clogged up with straw and it took me about a week to take all the man-hole covers off and give all the drains a good clean and to make sure they were running properly. After that I cleaned all the yard and burnt all the rubbish: now at last we had a clean start, which is the only way to do it.

You have an obligation in the food industry, to the people who buy your food, to have the standards right. I know this is old-fashioned and today people are more interested in how much profit can be made, how smart you can we be, but I didn't belong to that era. I belonged to the era where the customer comes first and the quality of the product comes first and the people who taught me my trade were good-hearted and their attitude was the most important thing. I was going to live by that and I felt I owed it to all the people who had taught me curing, including the Quakers in America who had shown me their way of life. I wanted to carry on their traditions, so my principal has always been: 'good food for good people'.

Back to my enterprise: that was the first stage completed and I knew in my mind's eye how I wanted to lay the little factory out. I'd worked in a lot of factories over the years so I knew which layout would be best. It is important to lay the factory out so that there is a continuous flow to the work line, with the product going on a one-way system.

I knew I wasn't making fast progress on my own and that I needed some help as it was important to set up the business and start earning a living. We decided to ring a local builder to see if I could get a helping hand. We happened to get in touch with a man called Sidney, who told me he was a builder. I explained that the main objective was to convert the dairy into a shop and the shippon into a factory.

He looked at me. 'That's going to be a big job,' he said.

I told him, 'We've both got to be stout-hearted: we can do it together.'

He looked at me strangely.

I said, 'Would you be interested?' and he said he would, explaining that he charged fifty pounds a day.

At that time fifty pounds a day was expensive, but I knew I had to get on with the job, so I agreed. Sidney went on to point out that he wanted paying fifty pound per day, at the end of every day, in cash. I was a bit taken aback but I agreed. We arranged to start the next day, and he arrived the following morning at eight o'clock.

We set to and were making good progress, when Trisha came down at nine o'clock with some bacon sandwiches. At lunch time she brought some lunch down, and at about four o'clock she brought a cup of tea and some cake. It was the custom where we came from that all visiting workmen were given their meals and you provided the food for them. It showed good faith.

We went on like this for a couple of weeks and when it came to repairing a wall, we were one brick short. Sidney said he had one at home and that he would bring it the next day. He did, and we worked all the next day until half past four. Sidney always stopped work at half past four; he changed his overalls and his shoes and by the time he had done that, it was dead on five o'clock: in fact you could set your watch by him.

I went up to the house and got his day's money and came down and gave it to him. As usual he methodically counted the money out, then said, 'The money's wrong.' I said I didn't think it was, but that he should count it again.

He said, 'I don't need to count it again: you haven't paid me for the brick.'

I asked him how much the brick was, and he said, 'Fifty pence.'

I said 'Alright, Sidney, I'll go up the house and fetch you fifty pence.'

I went up to the house and told Trisha that Sidney wanted fifty pence for the brick, but she said, 'No, don't give him fifty pence, give him a pound and that will make him happier.'

So she gave me a pound. I went down, gave him the pound and asked him if we were square.

'Oh yes,' he said, 'You've paid me for the day and the brick and I'll see you in the morning.'

But I said, 'No you won't Sidney, you won't seem me in the morning. I don't want you working with me any more. We have treated you with courtesy and kindness and fed you all day long and yet you have asked me for fifty pence for a brick. Life's too short to spend with people like you, so I would be very grateful if you don't come again.'

I think I had done the right thing as life is short: you don't want to spend it with people like that.

So I decided I would have to do most of the work myself. I planned to get the shop sorted out first and get the refrigerated cabinets in to make the shop look respectable. I bought the proper cabinets from a local dealer and the shop looked very nice when it was completed.

Next on the agenda were the curing house and the smoke house.

It is important when building a smoke house to put it in the right position to catch the right wind. You put balloons on canes in different positions over several days, to see which way the prevailing wind blows. Once you have done this you can site your 'fire box' so that it gets a constant draught. This is essential as there are no mechanical parts in a smoke house: it really is a case of earth, wind and fire.

A smoke house is basically is a fire box and a fire, with controls to harness the wind. I designed the smoke house on the back of a Kellogg's cornflake packet one night and I thought it was quite good.

I had already built one or two smoke houses before, but I wanted this to be a bit special. I designed all the flues and the chimneys and then went to see the local blacksmith, a jolly man who I don't think could quite weigh me up. I asked him if he had a piece of chalk and something to draw on. He said he had; he gave me the chalk and a piece of metal to draw on, and I drew the fire box. He looked at me strangely and asked me what it was. I told him it was a smoke house.

He said, 'We've never done one of those before', but I replied 'Well, that's what makes life interesting.'

He studied the plan carefully and told me he could make me one, which he did. The smoke house was designed with a long fire box and a dual chimney and I knew it was right. With smoke houses, the older they get, the better they get, because the tar on the wall gives the bacon added flavour. They aren't much use for the first few times so you need to smoke them three or four times to get a bit of seal on the walls. They seal themselves naturally with all the tar and after about a year they give the bacon a really nice flavour.

We gave our new smoke house a first run and it was a nice smoke and what you call a good 'puff', so all was going to plan.

We'd got the smoke house going and the shop was coming along. Next was the curing house and then we would be ready to launch. Ordinary concrete floors do not last in curing houses beyond five or six years as the salt breaks the concrete up. So I rang the local builders' merchant and ordered the special granite mixture. The agreement was I would pay on delivery, which I agreed to. The lorry arrived and I told the driver where I wanted it tipping, but he just sat there and said, 'You've agreed to pay on delivery.'

I said, 'Yes, I have agreed to that, but you haven't delivered it yet. Tip it up and I will pay you. At the moment it belongs to you. Tip it up and I will pay you.'

He did and I went indoors and Trisha gave me the cheque; I went out and gave it to him. I went back indoors and thought to myself: there is something wrong. Everytime I give a supplier this address, payment is the main priority.

I said to Trisha, 'This address is getting a very bad reaction. It's going to take some living down because every time we ring up for goods, the suppliers either want cash on delivery or cash up front. It will take years to establish that we are proper people. '

From then on, whenever we ordered more supplies I used to make a point of telling people we were the new owners of the property and in the fullness of time, that started to work.

My next task was to put the proper floor down. Floors in curing rooms are entirely different to any other floor because they have to be salt-resistant, and the secret to this is: once they are put down, you top them with granite. You have a small machine which spreads the granite mix, and you then smooth it out. The granite seems to glitter. The floors of a curing house are never level; they should slope towards one end, towards the drains. If you have them level, the water gathers in pools on the floor and can be a hazard, but with a slope, any surplus water runs down to the drains. Too

15

much wetness in the curing room encourages mould to grow and that is something in curing that you really do not need. I laid my granite floor and once completed, felt that was a good job done.

Trisha informed me we were spending a lot of money on the building, buying supplies and things, and that we were running out of money. We were half way there but the situation was serious, so I gave some thought on how we could raise some money. I wasn't happy to go to the bank; I've never liked banks and always steered clear of them. I prefer to generate my own money so the only solution this time was to sell a picture which I had inherited from my Grandmother.

I remember my Grandmother: she was a very mean lady. I used to be taken visiting as a child with my Father, and she used to grow strawberries, but I was only ever allowed to have one! Everything was done for economy. As life went on, she passed away. We went to the funeral and after it we all returned to her house and the solicitor was there. We all sat down, my cousins and myself, and Grandmother's Will was read. The lady who worked for my Grandmother came in and asked us all if we would like a drink. I had a coffee and some of the others had something stronger. The solicitor sat at the head of the table and proceeded to read the will. He read out that my cousin Wilton John Henry Walker had been left the bulk of the estate, the house, the furniture and any surplus money, and that Maynard Davies had been left a picture, '*Nymphs and Shepherds.*'

I thought to myself 'That's life!'

The lady came back in and asked if anyone would like any more drinks and I told her I would like a brandy as I thought I was going to receive a little more than a picture of nymphs and shepherds!

The picture duly arrived and it was so big it would not fit into the house so it ended up in the garage. I never really hung the picture but every time we moved house it came with us. It was surplus to requirements so the best thing to do was to sell it. I decided we would take it down to the local auctioneer. I rang and asked him if he would be interested, and he said he would like to see it. I loaded it into the cattle trailer as the picture was so large there was no other way of transporting it and took it to the auction house.

I took it to the sale room, the auctioneer came out and looked and said, 'We don't get many pictures of this size.'

He asked me how long we had had it, so I told him the story. He had a good look at it and told me it was of Italian origin, an unusual picture, and he would put it in the fine arts sale. I asked him how much it would fetch and he told me that, as it was a large picture, it would fetch between fifteen hundred to two thousand pounds and his commission was ten percent.

I agreed, so the picture was taken into the saleroom and did not look out of place amongst the other fine antiques. Duly the day of the sale came, and out of curiosity I decided to attend. The auctioneer announced the painting was the work of an Italian artist of the last century, albeit a little known one, but a picture in excellent condition (little did they all know it had been kept in a garage for the last few years!)

The bidding started at £500 and slowly worked its way up to £5,000, much to my surprise; maybe this art thing wasn't such a dead loss! I felt very pleased with the end result as it meant we could now finish the shop and the factory and buy the equipment we needed. I thought to myself, maybe it was fate that had given me the picture as my cousin who had inherited all my Grandmother's wealth had spent it all on fast women and slow horses and he had nothing left.

Everything was going smoothly, the shop was nearly finished, the cabinets were in, then the Health people wanted the walls lining. In the food industry, you line the walls with a polythene sheet and

I wanted to put a proper floor in the shop, just to give it a bit of a professional finish, so I looked in the local paper and found a tiler. I rung him up and told him I needed a tiled floor putting down. He agreed, came and looked at the job and told me he could do it. We agreed on the price, he arrived in due course and started to work.

On the first day he had not laid very many tiles and I thought to myself, 'There's something not quite right with this man.' The second day he came and he asked me for an advance on the money. I told him I could not do that, because I was in business the same as he was and I could not pay him until the job was done. I actually had my suspicions about him; I looked at the work and thought to myself, 'Well Maynard, you could do better than that when you are drunk!'

I let him carry on and on the fourth day I said to him, 'I don't want to be rude, but are you a professional tiler?'

I asked him where he had tiled before.

He put his hands up and admitted, 'I'm not really a tiler, I'm a long-distance coach driver but I decided I would have a go at this and change careers.'

I was astounded.

'That is dishonest; you've been advertising in the paper as a tiler and you aren't: it's totally wrong.'

He said he was a diabetic and that was the reason he had had to give up the coach driving, but I told him the hardest thing was to find craftsmen and honest people round here. So once again I had learnt a hard lesson. I paid him off then set to, finished the floor myself and wondered how I managed to pick them!

The time had come to buy the sausage equipment. This is a most important piece of machinery because sausage-making is a very skilled craft. A lot of newcomers to the industry think its just a matter of mixing water, bread, rusk and meat, salt and pepper and that's it. But that is incorrect: sausage making is an exacting art and the best sausage-makers are hard to find.

It is important to buy the proper equipment and one of the essential pieces of equipment is a bowl chopper which consists of a round bowl that turns at a certain speed, on a central shaft, with knives to chop the product. This piece of machinery not only mixes and chops the meat for the sausages, it can also be used as a mixer for other products such as pâtés and ingredients for your bacon curing. It is a very valuable piece of machinery to have. These days the bowl chopper has gone out of fashion.

The machine of today is called a mixer-grinder and it consists of a large arm rotating in a square container mixing all the ingredients together. I find that the mixer-grinder, as opposed to the traditional bowl-chopper, binds it all together so that it's more like cement than good sausage. The difference between the two is tremendous. You mix by experience and by eye with the bowl-chopper because all meat is different in texture.

The difference between traditional sausage-makers and some of the sausage-makers of today is tremendous: the traditional sausage-maker mixes all his own ingredients, and all his own seasonings. First, he separates the lean meat from the fat. When he is boning out, he takes out all the sinews and gristle. The seasonings are ground freshly on the day of production so he ends up with a superb quality sausage.

Today they buy sacks of ready-mixed seasonings with instructions on the label, put the meat through a mixer-grinder and add the seasonings. They can't see the meat separate from the fat because it all binds together in a thick paste.

I did not want to produce a product like this; I wanted to produce a high quality product for people to enjoy, for what is more

19

important than people's food? What is important is the satisfaction of producing good food: that should be payment enough. Today that is out of cog with the general thinking but I would rather think that way than produce something which I think is inferior.

Sometimes opportunity knocks very softly, and it did on this occasion. Our industry is a very small community in which everyone seems to know everyone else and the word had gone out that one of the best sausage-makers I had ever known, Edwin Hinton, was retiring. I knew he was a nice man and a great craftsman who produced exceptionally good sausages. He lived in a small village and had a shop and factory which produced a high quality sausage, so I found his number and decided to ring him.

I knew him slightly so I rang and told his wife who I was, and she recognised me. Edwin came on the phone and I said, 'I won't beat about the bush, I hear you are retiring.'

Edwin replied, 'I am glad you have come straight to the point and not beaten about the bush; that gives us a good platform to start from: what can I do for you, Maynard?'

I replied 'I would like to buy your beautiful machines from you.'

'Yes, they are nice, when would you like to come?'

'Tomorrow afternoon if that suits you,' I said.

'That's fine,' Edwin replied, 'I'll make sure the kettle is on.'

In our industry there were a few exceptional machine manufacturers and one of them was a firm called Barwins who had a factory just outside Manchester and the man who ran it was a complete eccentric, but he could make fabulous machines and at one time everyone in the industry bought these machines, as they lasted a lifetime.

When I was an apprentice, these were the machines used. They were so reliable. The bowl chopper rotated at exactly the right speed and the knives chopped at exactly the right time. The other machines he made, the mincers and the sausage extruders, were also right and you only bought these machines once, as they were expensive. These were the very machines Edwin had, so I hoped I would be lucky and be able to acquire them.

I duly arrived the following day. I went into the shop and his wife greeted me and told me to go on through to the factory. Edwin was there and what a place he worked in! Everything was sparkling clean and all the tools were hung up in line and everything looked like new. You judge a craftsman by how he keeps his tools and his factory, as this is reflected in the product.

We shook hands and I looked into his face and there were many years of work in that face and I think he had enjoyed every one of them. I felt very honoured that I had the chance to meet a great character, a gentleman amongst gentlemen and a craftsman. We talked, then we walked over to the machines.

I had worked on machines like these all my life; Edwin had had them from new and they still looked new, as they had been so well looked after. He asked me if I would like to look around the whole place. He told me he would be finishing at the end of the month, and then there would be a general sale.

I asked him if he had any regrets and he replied: 'No, in a life time I have had an abundance of happiness out of the job and I meet a lot of nice people. We might not have made a lot of money but we've enjoyed it. You know when the time has come to go: and this feels like the right moment for me to say goodbye.'

I knew the feeling as I had seen this with Thea, the old man who had originally hired me and taught me my profession all those years ago, so I tried to ease the pain for Edwin and asked about his seasonings. He told me he still made all his own and he would show me. He told me I was honoured as he did not show many people his seasoning room. I felt very privileged as we walked into the room: it was about eight foot by eight foot; in one corner was an old coffee grinder.

I said, 'I see you still grind all your herbs and seasonings with a coffee grinder.'

And he confirmed that he did.

'It must be a problem acquiring fresh herbs,' I said.

At that, Edwin asked me to follow him and he took me into the garden and there were all the herbs. 'We pick them fresh every day,' he explained. I looked round in amazement. There were great round circles with every kind of herb: coriander, sage, thyme, many herbs I had never seen before. We then went into the shed next door and there was a huge drying room with more herbs hanging up to dry. I felt like an unbeliever visiting the Pope!

I was very impressed with everything I saw around the factory. After a time he asked me if I would like a cup of tea. We went into the kitchen, which had a quarry tiled floor, a pine table and a big range with a kettle boiling merrily away.

I looked at this old kitchen and said to Edwin, 'You don't believe in modern appliances, do you Edwin?'

He smiled at me, 'No, we've had the old range ever since we came here: it's given me a lot of warmth and a lot of happiness and we've never had the desire to have a modern cooker.'

I felt privileged to be there. We sat down and had our tea and somehow it always tastes better brewed on one of these stoves.

'You produce a good quality sausage,' I told him.

He was very pleased with my observations and he told me he had only ever tried to produce the best. He told me he was pleased he was finishing, as the industry was changing, with lots of talk on how wonderful food is and everyone patting themselves on the back with awards and all the exhibitions on food but he said he was never a partaker of those.

He said, 'I will tell you a true tale on food exhibitions. I knew a man who used to enter the sausage exhibitions. He used to go to the local supermarket and buy sausages from a known maker and enter them as his own!'

I asked him, 'Did he ever win?'

And he replied, 'He came a good second!'

I asked him if he had ever sat on a judging committee and he replied, 'Just once. I had a grand invitation to be a judge, I arrived at the exhibition and we walked round, studying the sausages – the best presented, the best colour, the nicest looking sausage. Then we cooked them and tasted them, making sure the competitor's number was at the side of them. There were three of us judging and when asked my opinion, I said, "I think this man here". One of the other judges agreed with me: the taste was good, with good distribution and the man was a good sausage maker. The seasoning company offering the prize said, "No, he's disqualified: he hasn't got an account with us." I questioned this but the seasoning company stuck by their rules and disqualified him. So, Maynard, I never partook in anything like that again. The reward of a true sausage-maker, is people coming back every week to buy more and I felt I never needed the pieces of paper on the wall telling me how wonderful I was.'

I thought to myself: true wisdom. There's an old saying 'He who talks a lot says nothing.'

Edwin looked at me and said, 'You've come a long way; lets get down to business. You want the machines?'

'I would like them very much,' I replied.

He told me he had had them from new and they had been well looked after, they where the best sausage machines you could buy: simple and efficient. I told him I was familiar with them, and that I had worked with them before. He asked me if I wanted all three machines: the sausage extruder, the bowl chopper and the mincer. I told him I would want all three as they all lived together up till now and I would not part them. Edwin laughed.

He said, 'I've been thinking how much I would want for them now that I am retiring, and I think £4,000 is a fair deal.'

I looked at him, 'Edwin, they are worth £4,000. But I do not have £4,000 and if I did I would give it to you. All I can offer you is £2,000 and that is all I have.'

He sat deep in thought, then leant across to me and offered me his hand.

23

'Shake on it and we'll call that a deal,' he said.

I thought to myself that there are very few men of this calibre about: a man who had worked all his life, producing good food for not a lot of payment, but he got his reward in pleasure for what he did, and that was the greater payment. So we agreed on £2,000 and the deal was struck and I was nearly ready to open the shop.

We decided then to set a date for opening so it was all hands to the pumps. I started to prepare the pigs, made up all the brines and gathered together the ingredients. I intended to produce the best bacon that I could. Not only would I use all my skill but I would put plenty of love into my product too.

I knew I wanted the best ingredients I could find for the production of the bacon. It would have been easy to buy the ready-made mix from the butchers supply merchants and just stir it in and hope for the best, but for me that was not good enough.

I was a traditional bacon curer and I was going to give people the bacon they had paid for and they deserved, so I went to Grace's in Liverpool who I thought were the best people in the business for the ingredients. I had dealt with them before and they knew their stuff. They were an old established firm in the industry and were very knowledgeable. They supplied all the salts, spices and sugar to the industry and it was a pleasure to deal with them. They were professionals amongst professionals and in my opinion that is the magic ingredient which is lacking today.

In bacon curing, different salts give different tastes. Whether rock salt, sea salt, lump salt, dairy salt, bay or roman salt; they all give a different flavour to the bacon. The skill is in mixing the right sugar with the right salt and they are all different grades and impurities to give the desired taste. This company stocked all the main sugars: muscovados light and dark; molasses; all the treacles and all the herbs which give bacon its distinctive flavour. I think

24

if you start in the beginning aiming for quality and using the best ingredients, that must be the road to success and that is the way I wanted to go.

One evening when the daily toil was done I decided to map out the recipes for my company. This would give us a platform to start from to produce a good quality product. When I served my apprenticeship at Thea's, we had a lot of old recipes, but many of them did not make commercial sense today as they were out of favour. Basically speaking, we cannot eat as much salt in our bacon as our ancestors could. Salt in those days was only added to cheese, butter, bacon and very few other products, unlike today when salt is added to all processed foods, especially fast and convenience foods which are on the shelves in all the supermarkets.

So I had to make a new beginning and we had to adapt the recipes to produce a new bacon that was palatable for today's needs. I had a lovely gift given to me by my apprentice-master Thea, which was his original recipe book and I tended to study it a lot over the years as it brought back memories of days gone by and wonderful times. You could say it was my bible.

Reading it, my mind cast back to my days at Thea's and the old seasoning room, a large room with shelves all round with all the seasonings marked up in containers. In the centre of the room was a huge mixing bowl, something like an old butter churn. All the ingredients had gone into this and then it was sealed up and turned round and that was how we used to mix all the different seasonings. We used different combinations for everything in the factory from the sausages, bacon, cooked meats, polony, black puddings, pork pies, everything we did: that seasoning room was the hub and that was where the skill was. Thea's knowledge was boundless and I was very privileged to work with him. He used to tell me to gather all the ingredients and weigh them.

I used to say, 'Well Mr Thea, its six ounces,' and he would reply 'How do you know its six ounces, without looking at the book: it's a fortnight since we last weighed these ingredients?'

25

'Well,' I would reply. 'It was six ounces the last time I weighed these ingredients. I have difficulty reading, so I have to remember all the recipes and for anything important, I have a photographic memory.' (Dyslexia was not recognised at this time)

He scratched his head, 'One gift for another; a gift for a gift.'

These memories made me determined to try and put a little bit of the past into the future, to put back a little bit of the skill that was being lost. I started copying the recipes down: the recipes of my apprenticeship down and the first one I decided to do involved treacle. I came from the North Staffordshire borders so I named it the Staffordshire Black; and that was the first recipe that I did. I remembered it clearly, wrote it all down and measured it all out.

I was well underway now. Next I did the Traditional Cure, which follows the dry salted method, but isn't overly salty; then I made up the Old Fashioned Cure which is a saltier verson of the Traditional, and is popular because it takes many of its fans back to their childhoods.

Then I did the Old Fashioned, the Colonial which is Canadian-style and cured in maple-syrup, and the Sweet which hales from Cheshire. Then the Welsh which is a mild bacon with a good layer of fat and the English Gold which is a sweet, lean bacon from Norfolk and then a honey-cured one.

Other bacons I recorded included a Northumbrian Full Flavour, a spiced Devonshire bacon, the York bacon which is cured with a particular natural salt. Also the Wiltshire bacon which was invented by the Harris Bacon Company of Calne in the C18th and the Shropshire bacon which is cured with herbs and spices and, curiously, was a particular favourite in the taverns of London. Tamworth bacon is made from that breed of pig of the same name and has a touch of ginger in the cure. I also recorded a farmhouse bacon cured with paprika, and the Harvest bacon which dates from Roman times and of course the Somerset bacon which has the local ingredients of cider and honey to give it its wonderful flavour.

Spiced London Bacon

This is a popular bacon we used to supply to the London market. It is delicious, with a distinct taste: lean and cut very thinly.

Ingredients
30lbs bay salt
15 gallons of water
6lbs light muscovado sugar
4oz pepper

6oz salt petre
6oz ground allspice
4oz dried coriander

Method
Boil the water and leave to cool. Add the bay salt and stir well. Dissolve the salt petre in the correct way. Wait until cooled, then add it to the brine, stir well and leave for 24 hours.
Boil the sugar in water, leave to cool. Add to the brine. Now put the lean bacon in brine. Leave the middle cuts in it for five days. They can also be pumped at 65/70 density, higher than normal as this bacon is cut very thinly. Remove bacon from brine, wash off and hang up to dry. When thoroughly dry, remove the rind and surplus fat with a sharp knife. Take a meat hammer and level. Grind pepper, allspice and coriander and sprinkle over the bacon. Take the thick end of the cut, fold over and roll up. Starting in the middle, string along tightly, place in a smoking net and hang to mature for about a week at 60-70F.
Then either leave green (unsmoked) or lightly smoke with beech and barley straw for about a day and a half and leave in smoke-kiln to cool.

I decided as the smoke oven was running well that I would do different types of smoke. I did a smoke with an apple wood which was nice, then a hickory, a juniper and naturally I did the oak which is the traditional one.

I had laid the foundations by putting all the recipes down and I now had a full recipe book and I thought that was the beginning of the beginning.

One of my favourite old recipes, which was full of memories for me, was the old Quakers Settlers recipe and I was a bit dubious about producing this because it was a very strong, salty bacon which the pioneers to America took along the plains. They needed a strong salty bacon as it had to last them weeks and weeks on their journey in wagons; it was cured to last a long time and its main purpose was to give their beans flavour.

The recipe had been given to me by an old Quaker lady in America many years ago and I was unsure whether it would go down well in modern Britain, but I decided to do it all the same. I did it exactly as she had written it down in her wonderful hand and the first production, believe me or not, was like eating with God and it was a compliment to her as she had given it to me as a gift. The Settlers bacon became very popular with some of my customers. These recipes were unique to us and not produced from a ready-mix bag, and we were supplying people with a proper, individual bacon.

Everything was going to plan and my youngest daughter Anna decided to join me in the business, so I felt very grateful and happy about that. As the big day for opening was nearing, we decided we would have a dummy run so Trisha and Anna went to one side of the counter and I went to the other and we practised taking the money and handling the produce. We had not done it for a while and we wanted to look professional.

Finally, we set the shop out: it looked all nice and clean, we put up a big sign saying, 'Welcome' and with a lot of spirit and a great deal of hope, the day of opening finally arrived.

People used to ask me how long had I been in the bacon trade and I used to tell them, 'All of my life.' Then their response would often be: 'I bet you know everything there is to know about bacon.' and I used to reply, 'No, the old curers used to have a saying: *When you think you know it all, that's the time to retire.*'

And I didn't know it all. I was never confident, always nervous, and I always had to prove myself - so every time I cured I had to give it one hundred and ten percent and I think that was a little bit of the secret why people said it was good.

We decided to open on the Tuesday and everything was ready; it was like D-Day. We were all up early, had an early breakfast, went down to the shop and set everything out. 'Setting out' means cutting all the bacon, setting it all out on trays, then arranging the display in the counter. I had cured about five different varieties of bacon: the Tamworth, Sweet, Old Colonial, Apple Wood smoked and some Oak smoked. I had a good display.

By half past eight everything was ready and I thought, 'Right, Maynard, this is a new beginning.'

Nine o'clock came and we had had no customers, then ten o'clock and still nobody.

I thought to myself, 'This isn't a very good start.'

At about half past twelve a lady came and she said 'Is this all your own bacon and is it any good?'

I replied 'Yes, it is all my own bacon but you will have to be the judge if it's any good.'

The lady laughed. 'Do you think you will be able to make a living round here selling bacon?'

'I shall try my very best,' I replied.

She said 'I wonder if I dare buy some?'

I said, 'I don't think it takes a lot of courage and I can guarantee it will be right.'

The lady smiled and I chose some of the traditional cure, wrapped it and handed it to her.

'How much do I owe you?' she asked.

'You owe me nothing.'

The lady looked at me, 'This is a strange shop; why don't I owe you anything?'

I smiled at her.

'You are my first customer and I am a bit superstitious, so I am going to give you that, for as bit of luck.'

She looked at me. 'I think you will do well, and good luck to you.'

'Thank you,' I replied.

We traded slowly and steadily all day and we did not take a lot of money – it's strange how things stick in your mind and I remember to this day that we took £35. We had worked all day, the two of us, Anna and myself, for £35, so I thought to myself: this looks like bread and margarine all week but I did not let that deter me and I determined to come back fighting the next day.

CHAPTER 3
Cheese, Bacon and Bureaucracy

The next day came, we set all the bacon out in the refrigerated counters and waited for the customers. It was as if heaven had smiled on us: we had a lot of customers, perhaps the word had gone around we were giving stuff away! I was very pleased the business was taking off and people were saying it was a good product and that was payment enough if I was making a living and selling good food.

Things went from strength to strength and it came to the stage we were working long hours to produce the bacon and the business started to grow like Topsy. I thought to myself, all the hard days when I was an apprentice and all the things I learnt, were well worth it. I was the last to have served a traditional bacon curing apprenticeship in England and now I was giving people the good food they deserved.

As the business started to grow, the time came to employ some help. We took on a nice young lady called Mary and she was a pleasant girl, very clean, very co-operative, very quiet, so more-or-less you had your own company with her.

I remember one day we were all working, Anna, Mary and myself, when Trisha came in and said the house was a bit shabby and we could do with the dining room being painted. Mary piped up and said if we looked in the local paper we would find a decorator in that. We bought the local paper and true enough, we found a handyman to do the painting. He seemed pleasant enough so I thought we might as well set him to do the painting.

He came about ten o'clock the following morning and started to paint. I went up to have a look and was very impressed and thought how fast he was: he had done two and half walls before half past eleven. I went back to Trisha and Anna and told them he was the fastest man with a paintbrush I had ever seen in my life and there was not one bit of mess around him.

'As daft as it may seem, he will be finished by half-past three but I will still pay him a full day's wages.'

I said to Anna, 'I think we've dropped on our feet here.'

We had a fair day in the factory and about six o'clock we went up to the house to dine. After we had eaten Trisha announced she wanted the furniture moved in the dining room: you know what women are like, not all of them, but the majority like to move furniture about. The dining room looked perfectly nice after its coat of paint but Trisha wanted the dresser moving.

I said, 'Not tonight,' but Trisha was adamant that it wouldn't take a minute.

So we took all the ornaments off the dresser and we all pushed the dresser and we could not believe it – he had painted round the dresser! We all stood looking at it, unable to speak.

Finally I said, 'I don't believe this, this is crazy.'

Trisha replied, 'Well, believe it.'

So we went round the whole room and discovered he had painted round everything: pictures, the corner cupboards, he had a bit of talent in his hand as he had even gone round the Grandfather clock and really Rubens or one of the Master painters could not have done it better; he had painted round that clock like a professional artist! We had to admire his skill.

The business was really going very well, we were busy and we employed three ladies, as well as a part-time man to do the heavy work. His name was Harold and I used to show him what to do. One of his jobs was to load the smoke oven as we used to smoke twice a week. The smoke industry was going very well at that time; we had a properly designed smoke house, a natural oven which produced very good bacon and I was proud of it. I had put a lot into that smoke house and I thought we were being repaid a little bit.

The spin-offs of smoking are that people want you to smoke different things. I used to smoke things like cheese, salt, garlic but I never smoked fish because that would taint the oven. You need a separate oven to smoke fish, but I would smoke anything else people wanted.

On occasion a very good cheese maker used to get in touch with us. He used to do a lot of exhibitions and show his cheeses, and one day he asked us to smoke some of his prize cheeses for the Nantwich Cheese Fair. He brought along two lovely great cheeses for us to smoke. I told Harold to put them on the bottom rack and give them a light smoke.

How you smoke cheeses is by laying them on a small mesh table. The fine mesh is made of stainless steel, which allows the smoke to penetrate gently and evenly. The low table is placed on the floor of the smoke house where they don't get too much smoke as it rises. To achieve an even lighter smoke, the cheese can be wrapped in a muslin cloth which has been soaked in cider vinegar to prevent it from sticking to the cheese.

You load the rest of the smoke oven with your hams and bacon. When the smoke is over, the smokehouse door is opened to let the smoke settle; it is a lovely smell, and when the smoke finally clears you can see what kind of a job you have done just by looking at the smoke hanging in the oven whether it is right or wrong; you can tell by experience and there is no substitute for experience.

So we unloaded the oven and when I saw cheese I was horrified: I could not believe my eyes. Harold had hung the cheese up just like the bacon and with heat of the oven the cheese had curled round like a horse-shoe. It took me completely aback: how could he have done that! I had to go and sit down. I had never seen anything like it in my life and I thought to myself, 'What a mess!'

There wasn't sufficient time to smoke any more as I knew he was picking up that day to go to the show, to have his cheese judged the day after. Time dragged slowly on until he rang me up and asked me how the cheese was. I asked him if he was sitting down; he went very quiet so I asked him if he wanted the good news or the bad news first.

He replied, 'I'll have the bad news first.'

I said, 'Your cheeses have unfortunately been hung up instead of being smoked laying down and unfortunately they are horse-shoe shaped.'

The line went very quiet for a minute and a strangled voice said, 'A horse-shoe shape?'

'Yes,' I replied, 'Both cheeses. It's our fault and there will be no charge and we will do your next smoke for nothing.'

He said, 'But there will not be time. I should be picking them up today for showing tomorrow.'

I apologised profusely: I had never felt so embarrassed in all my life. He picked the cheeses up and he still entered them in the show as he had no alternative: he was already booked to enter. He got third prize for smoked cheese and in another section he won first prize for Ingenuity because the judges had never seen a cheese like that before and neither had I! Often we would meet after this episode, and we would laugh over it.

We used to have visitors from the local Health Environment offices and, generally speaking, they were nice people, but like all organisations there was the occasional person who took themselves too seriously.

Everything was running smoothly when one morning the

Environmental Officer turned up to do the annual check. It was a young woman I had not seen before. She wanted a tour of the factory, so we started there and she asked me where I kept the medical kits. I pointed them out in the shop with the red cross on; there was also another one in the factory, also with a red cross on.

She was very pleased and said 'That's very good.' She then appeared deep in thought and asked me how I dealt with the rodent problem.

I replied, 'Oh, we have our own Rodent Officer.'

She looked surprised at this.

'You have your own Rodent Officer?'

'Yes, 'I said, 'He's very efficient and his name is Jack Scrapp.' I didn't like to tell her that at that precise moment he was right behind her, sprucing himself up for the day, having just enjoyed a large saucer of milk and a plate of fish!

Next she wanted to inspect all the machines. In the food industry, the custom is that at the end of each day, the machines are scrubbed down and then covered over with a cotton cloth to keep all the dust out. She was determined in her own mind that we were hiding something sinister under the cloths and wanted all the cloths off so she could inspect.

So we started with the bowl chopper; the cloth was removed and she had a good look, then on to the sausage extruder, cloth off, inspected and then onto the meat grinder, the slicing machine and she must have decided after that she had seen enough. We then went outside to the smoke house: she went inside, paying particular attention the walls.

'When do clean the smoke oven out?'

'We never clean the smoke oven out.'

She looked at me, 'But you must. You must wash the walls down.'

I replied, 'You can't wipe the walls down, it's an impossibility, that is how a smoke oven is. That is how you get the taste; that is part of our heritage.'

35

She still persisted, 'Surely you can clean them down?'

I was beginning to lose patience.

'No, you can't clean them down.'

I thought to myself: 'She's got a piece of paper that says she is clever and that piece of paper is telling lies!'

We went back to the factory and she said I would have to get rid of the wooden cutting boards because they were not hygienic enough for the EEC's rules and that I would have to use plastic cutting boards. The following day I went out and purchased the plastic cutting boards.

After inspecting all these things she seemed satisfied and told me she would write to me. As she went through the door I thanked her for her visit. I thought to myself, she had not made up her mind whether I am a clever man or a complete idiot, but I left that decision to her!

Two days later she rang me to tell me the policy had been changed on the cutting boards and that we could have the wooden cutting boards back because they found if you used the plastic cutting boards they were less hygienic and if they chipped and it got into any of the products, then into the body, then the body could not digest plastic as well as wood, so it was recommended we used the wood again!

I thought to myself 'What a crazy world this is!'

In the bacon industry you have to have an especially good supply of oak and oak sawdust, and it is essential you acquire the right kind of oak, as each oak has its own flavour. Welsh oak is traditionally a very bitter oak and it isn't good for smoking as it gives a bitter bacon. The amateurs amongst us don't know these things and they think sawdust is sawdust but it isn't. Spanish oak is another one that does not produce a good smoke and most of the conifers are not very good either.

It is essential to have a working knowledge of the best materials to smoke with and I acquired my knowledge from one of the old smokers called Bernard who worked at Thea's and he knew his woods. He could put his hand in a bag of sawdust and he could tell you which wood was in it and if it was a mixed bag of woods he knew. That was what you call skill and dedication and unfortunately this is one of the skills that has been lost in our industry today. Now it is often a bit 'cross your fingers and hope for the best,' but when I went to this small wood-working company and explained to the man, of course he did not think selling me a bag of sawdust was of great interest.

When I mentioned, instead of paying him in English currency, that I would pay him in good quality bacon and hams, his whole attitude changed! He said he would be very pleased to have my custom and I thought to myself: doesn't bacon grease the wheels of industry!

So he took me down to the wood-sheds and introduced me to a marvellous man called John who knew his woods and he said, 'What do you want, young man?'

I thought that was a lovely compliment as I wasn't a young man any more. I replied, 'I'm a smoker by trade, a traditional bacon smoker and it is essential I have the right kind of sawdust to smoke the bacon, as different woods give different tastes and odours.'

He agreed with me, 'There's a lot of different resins in the different woods,' and I knew by that, that he knew his woods. He asked me which woods I was looking for and I told him a hard oak and a gentle beech. He asked me if I wanted an English beech and I said that was correct. I didn't want any sycamore and he agreed with me as he said it burnt too fast.

He asked if I wanted anything else in particular, and I replied, 'I do like, to be honest, an apple wood but I suppose that is an impossibility to acquire.'

He smiled. 'We do buy a lot of apple trees from Worcester to make special furniture and sometimes we have a lot over. I will bag it for you.'

I thought to myself, 'I've struck gold this morning,' and it was indeed a great partnership we had entered.

As the years went on I remained good friends with this wood yard. It was a nice exchange; I took him the bacon he enjoyed and he gave me good sawdust.

On one occasion I arrived for some sawdust with my old truck and John and I started talking. He was also a breeder of shire horses. In fact, horses were his life and he could spend all day talking about his horses. One of his horses was called Samson and he always used to tell me how Samson was.

One day I committed a cardinal sin. I knew Samson had won a good prize at the County show so I said to him, 'I suppose Samson will be at the next show this weekend with all the rosettes on his head collar.'

As soon as I said it I knew I had done something silly. He went very quiet and he looked at me.

'Maynard, Samson doesn't need all the rosettes on his head collar, he knows he's the best, the judge knows he's the best and I know he's the best, and it's like everything else, Maynard, you don't have to tell a beautiful woman she's beautiful – she knows.'

I thought I was well put in my place and it had been well-deserved. I always remember that lesson and it was a good lesson to learn.

One of the three girls who worked for us had a very strange surname which she wanted to change and she kept asking me how she could go about it. I knew there was a solicitor who came in to the shop regularly and I thought I would ask him.

One day the girls came to me and told me he had come into the shop, so I went and asked him if he could give us a bit of advice on the best way to change her name. He said it was easy: just go along to her solicitor and it could be changed by deed poll.

I served him, added the bill up and, from memory, I think it came to about three pounds. He put a pound on the counter and he said, 'I've knocked the two pounds off for advice.'

I thought to myself: 'True to form.'

When you are dealing with customers every day, you meet all sorts of people. I recall one, a lovely lady - but unusual. She came in one day and she said to me, 'Maynard, I know you like old things and I wondered if you would be interested in buying a clock off me?' I said I would if it was a nice clock and she it was, and asked me if I would like to come and look at it.

I said I would, but she lived in such an out-of-the-way, rural area, I told her I would never find it so she said she would come and pick me up after the shop had closed.

She duly came and I followed her in my truck. It was good job she was leading as it was lane after lane and I would never have found it. We eventually arrived and it was a lovely old cottage with low ceilings, an inglenook fireplace and some lovely furniture. She pointed out the clock and I put it on the table to have a good look.

It was an exceptional piece and I asked her how much she wanted for it. She told me she was asking £350.

I said to her, 'You do realise that clock is worth more than £350, as it is Gustavo Becker clock, and it is worth a lot more than that.'

She looked at me.

'Yes, I do realise that, but I want £350 for it.'

So I paid her and took it home. I put it on the mantelpiece and it looked grand. It chimed and struck the quarter hours and I thought how lucky I had been. The clock was beautiful, and the next time she came in I asked her where the clock had come from.

She said, 'Maynard I will tell you the truth. My husband was in the war and when they liberated Berlin they went into this large villa and took the clock. I have got rid of everything that belonged to him and that clock was the last thing that belonged to him; he was an utter b****** to me and I thought that the best thing to do was to get rid of it.'

I thought to myself, 'Well, that's life.'

We were beginning to integrate into the local society. There was an old farmer called Jethro who used to come round and have a chat and he used to tell me how everything used to be done on the farm by the horses, and that intrigued me. He told me they had 30 horses on the farm to do the work. You meet some people in life and they leave you with something; and I used to be enthralled by his stories.

He was in his eighties and in time he passed away and I was told the date for the funeral and invited to attend. The day came, we arrived at Jethro's farm. It was up a winding track and the house was like a lot of farm houses; it had been neglected and outside the house there was a lot of farm machinery and you could see there had been a lifetime's work done there.

There were a lot of people who had come to pay their respects and a lot of run-down vehicles, as farmers always buy the best cattle and the worst cars. Like most farming communities, it takes an occasion like a funeral to bring everyone together.

Country funerals are quite different to town funerals: a good meal is put up before the funeral, with plenty to drink as people had travelled from quite a way. On this occasion the vicar who took the service was an ex-army padre, quite a character, and he was partial to a snifter. He took more than one or two at the pre-funeral lunch and he was slightly the worse for wear.

The coffin was loaded onto the hearse and people stood respectfully by while it drove away to the church. The rest of the funeral party followed onto the church and the service progressed normally with plenty of hymn singing and the sermon. I could see the vicar was a little the worse for wear, and seemed a bit unsteady on his feet, so I was hoping the occasion would go off with no untoward happenings. Then the coffin-bearers carried the coffin out to the side of the grave with the entire congregation following and we all stood round and the vicar said a few prayers. The bearers slowly released the ropes and the coffin began its descent into the grave.

The vicar standing at the head, slightly unsteady on his feet, intoned, 'Ashes to ashes, dust to dust,' and threw the soil onto the coffin but unfortunately he followed the soil and fell headlong into the grave on top of the coffin.

As you can imagine, the situation was very awkward and it was one of those occasions in life when you do not know whether to laugh or cry. I managed neither and I kept a straight face, although I thought to myself, 'Jethro would have enjoyed every minute of this!'

After a few minutes of panic, one of the coffin-bearers got one of the ropes and threw it down to the vicar who tied the rope around his waist and was hauled up, dislodging more soil onto himself on the way and it was quite an effort to get him out. He was covered in soil but he dusted himself off and finished the funeral.

As I walked back to my car I thought, 'That was what you call an unusual send off!'

CHAPTER 4
The Past and the Present

When I was deciding on which of my original recipes to use and which ones would be popular with the people of Shropshire, it reminded me of days gone by and of the time when I was about eighteen years old and Thea was taken ill. I was informed by the lady who did the books in the office that Thea would not be in for a few days. His illness was self-inflicted because he did rather like the Chivas Regal whisky and we all knew in the factory that it was self-inflicted and that he would need a couple of weeks drying out!

The message came from Thea that I was to do all the seasonings and to keep the factory going until he returned. At my age it was a bit of a daunting position to be in. I did not have the knowledge of old Thea by any means, as he was very talented man. He never showed off; he just did it, but would always show you if you asked and that is the sign of a good craftsman. Poor craftsmen never pass on their knowledge because if they did tell you, they would not have anything left!

The seasoning room was the hub of the factory as this supplied all the seasonings to the sausage house, the pie department, the cooked meat department and the curing house, so the seasonings had to be right. The food in those days was of a high standard and if you did not give people quality, you did not remain in business very long. Today you can buy food which is very expensive and poor quality, and it amazes me how people are prepared to pay such high prices for poor-quality food.

Anyway, I stood in the seasonings department and decided to get to work. It was quite a job to get everything right: I did all the sausage seasonings, the different pie seasonings; we used to do Grosvenor pies, a Melton Mowbray-type pie, a plain pork pie and we used to do a pie called a Wagon Wheel which was like a pork pie but very large – they have now gone out of fashion and I haven't seen one for years.

There was a different type of sausage meat for sausage rolls. For them, a trade secret is to add corn flour to the mix which stops the sausage meat from shrinking inside the pastry. It was quite a responsibility to make sure the bakery was running right and when the bakery was very busy, I was seconded to this department to help the ladies make the pies.

There was an old Peel Oven, which was even in those days well out-of-date; it had an arch door and everything went in, fed from the back. When I was seconded in to help the ladies, they taught me how to make the pastry, and all the roses for the decoration, different designs for the tops of different pies, and I received great satisfaction from learning those skills. It was all part of my training.

The next place I had to prepare for was the sausage house, for which they have four or five different seasonings. All the mixings had to be weighed out the night before, ready for an early start the following morning. The cooked meat department also had all the different seasonings for their pâtés.

Whilst Thea was recuperating, I had to mix all the recipes for each one: three different types of pâtés, a liver sausage, a black pudding, and stuffings, so it was quite a large job.

Finally there were also all the mixings for the curing house. I got myself into a nice routine and with the arrogance of youth and all my inexperience, I thought I would alter some of the seasonings as I decided some of the recipes were a bit dull so I altered the mixings for the sausage-making.

Fortunately for me, the sausage trade during that fortnight went up! I also did the same for the bacon trade and some of the hams I thought where a bit dull, so I geed them up a bit too.

Hilda in the office said, 'You can't be doing too bad, Maynard, the sales figures have gone up and the product is lovely.'

I thought to myself, 'Fools luck!'

As Thea remained on the sick list and not at work, the responsibility had to be shared and I had taken over the seasoning room and I did the best I could. The next job that Thea did was to buy in the livestock.

Nobody wanted this job, going to the auctions, as it meant early mornings and late evenings, so as usual I drew the short straw. I was instructed to go and buy the pigs, since Thea had passed on a message to say I had enough experience to do it because I had been an apt pupil and I had to start some time. I went to see Hilda in the office and she issued me with the company cheque book and the cards used for the marking when you had bought the pigs. Funnily enough, old Thea's stick was in the corner and she presented me with that too, which seemed awfully funny to me.

So the morning for the market came, I turned out early and was at the factory at about quarter-past six. I went to the old Trojan truck and it had seen better days and not a lot of love had been spent on it. I put the starting handle in and gave it a turn and she started straight-away. I then went back to the hay store and put some straw on the wagon so that the pigs I brought back would be comfortable.

As I closed the doors at the back, I knew everything depended on me. I was eighteen years-old and to be very honest, I was very nervous. I'd got the cheque book, Thea's stick in my hand and not a lot of experience, yet it depended on me to buy the right stuff at the right price. So I set off for the livestock market.

I arrived at the market and parked, went to the auctioneer's office and knocked at the little window. The auctioneer asked me what I wanted and I explained that Mr Thea was ill and I would be taking his place.

He looked at me and smiled, 'So you will be buying the pigs – you're the man!'

I said, 'Hopefully,' and the auctioneer nodded.

'Have no fear, you've been with Thea a long time and you could not have had a better teacher. You will soon get into it and enjoy the experience.'

I went down to the pig area where the pigs were; in fact I mimicked Thea a bit in the way he used to walk down the alley where the pigs were and tap the sides of the pigs with his stick. I used to watch him and if the pigs were docile, he passed them over on the grounds that they were probably not healthy. He used to walk up the alley and back and he knew which he was going to buy. He knew by looking at them exactly how much they would weigh dead, so with each one, he knew how much he was going to pay for it. The skill is not here today, unfortunately, but I was lucky enough to be at the end of an era where these skills were still taught.

So I did the walk and I mimicked Old Thea. I walked the aisle the first time and tapped my stick, and some of the pigs grunted. I thought to myself, 'That's a good gilt,' and I looked at some and they had not been castrated and I knew they would not be any good. I looked at their plonkers like Thea used to do; you could tell what they had been fed on by doing this, which ones had been fed on good corn, and which ones had been feed on cheap food.

I think I must have done the walk ten times because I was so unsure of myself but eventually I had in my mind's eye the ones I wanted, so on the back of the cheque book, I wrote the numbers down of each of their pens.

Altogether we needed forty pigs to keep the factory going and

I was also conscious I had a lot of people's jobs dependent on me. There was not a lot of spare money in our industry at that time and I knew I had to do this right.

There were some of the regular buyers there and we all knew each other. They watched me and gave me words of encouragement which made me feel very proud.

At the livestock market there is a tradition: when it is ready to start, a big bell is rung. It reminded me of a prize fight, with everyone ready to go. The auctioneer arrived, and he always used to wear a hat which I admired. The time had come and all the buyers drew round. It seemed odd that old Thea was not there, as he had been doing this job for the last thirty years but on this occasion, he wasn't there.

The auctioneer began: 'This is a pen of six gilts; first bid please gentlemen.'

I put my hand up, the auctioneer looked round, but nobody seemed to be bidding.

One of the buyers said to me, 'Maynard, this is your first and last concession you will have here: as you are the new buyer for Thea's, we will give you the first pen with no competition.'

I was very taken aback. I looked at the auctioneer and he smiled; it was great to be a gentleman amongst gentlemen, and I have never forgotten that moment. The auction continued on and I had bought my first lot of pigs. As the buying progressed, I began to feel more confident and I walked with a bit of a swagger and I thought to myself, 'I'm a buyer!'

The auction finally finished and as usual I brought the truck over and waited my turn to load up. I was very proud of myself as I knew they were good pigs. They all went into the truck like they were going on their holidays, and I had put straw down for them, just like a slumber-down mattress! I also thought they needed respect and kindness, the same as any other living creature. It's like a gardener tending a tree that bears fruit: a little bit of happiness comes back.

As usual, I then fetched my vinegar and doused all the pigs so they would all smell the same and be one happy family – this is how to stop them fighting. I then drove the truck back to the parking area and went to pay my bill.

We were all standing there in the queue to pay, and one or two of the buyers asked me when Thea would be back: one wit commented that Thea could be his own Doctor and be back when he decided. They all knew Thea liked a drop of whiskey!

My turn came and I asked the lady if she could fill in the cheque for me, as I couldn't write. She was happy to do that for me. Thea had already signed it and I was given the receipt. So that was my first experience in the world of being a buyer and as I was now a fully bona fide buyer for Thea and company, I decided I would forego my usual fish and chip supper and go to the *Sutherland Arms*, Thea's usual haunt, to have my lunch.

I walked in with a swagger and the barmen said to me, 'Thea not with you?'

'No,' I replied, 'He's not very well.'

The barmen winked at me and asked me what I would like.

'A ploughman's lunch, please,' I replied, 'And a lemonade.'

I sat down. It was very busy as it was market day. The young lady brought me my lunch and when I asked what I owed, she replied, 'It's on the house with our compliments.'

I thought to myself: 'What a grand thought!'

And what a marvellous experience it had all been, my first time as a buyer of pigs: I suppose it was a landmark in my life and I thought I had done quite well, and I was very pleased.

Soon after Mr Thea had made a full recovery, he came up to the seasoning room one day and asked me how things were. He always had a polite way and he was always direct which I liked: you knew where you were with him.

He said, 'You've done a good job but you've altered some of the mixings haven't you?'

Yes,' I replied, 'I have altered some of the mixings.'

He looked hard and long at me.

'I'm going to do two things: I'm going to give you a rise and the bacon you've altered I'm going to call Maynard's bacon.'

I was taken aback.

'I don't want you to do that, Mr Thea. I will accept the rise but it's been a communal effort while you haven't been very well and to single me out would put me in an embarrassing position. I would like to accept the rise if you think I am worth it, if that is alright with you.'

He looked at me with a twinkle in his eyes.

'I think that is a very wise decision, Maynard,' he said, and we settled on that.

Bacon curing is a very small world but it is divided up into many sections. Now with my own established business, I would be asked on occasion by fellow colleagues in the industry when they had problems, if I would come out and give my opinion and luckily sometimes I could spot the problem and put it right.

I remember on one occasion a man whose niche in the industry was boiled and roast hams, rang me up. He did all varieties and he also used to buy the legs of pork and convert them into hams which is a very simple operation.

When he rang me up and asked for help, I told him I would be down in a couple of days, as I was tied up. True to my word, I turned up a couple of days later and he took me to the factory asking me if I would like to look at the production line, explaining that a black mark was appearing on some of the hams: not in all of them, but enough to be causing a problem. He provided me with 'whites' and I always take my own hat. The hams I saw were coming down the conveyor belt and, true enough, every fourth or fifth ham had a mark on it.

I could not understand it. We inspected all the hams and I looked at the finished product, and as the hams had cooled down you could clearly see the black mark. So I advised him to stop production and try to sort it out, as he was just losing money, losing one ham in five.

First off, I told him to send the line staff for a tea break while we looked to see if we could find a solution. I then asked him what he had done with the marked hams and he said they had cut the mark out but I told him he should not do that as he was just masking the problem and the cause needed to be found.

We went back into the factory and stopped production. We went into the cutting room and I said to the head butcher, 'Give me five hams,' and we picked five out and numbered them, then put them onto the conveyor belt.

As the five came down, I walked to the other end of the factory and as they came down the belt, sure enough one had a mark on it, a black mark. I thought to myself, 'This is crazy, why is there a black mark?'

I went back to the start of the conveyor belt and I asked for one ham; it was duly marked number one. I followed it all the way along the conveyor belt, which went up into a chute and dropped onto another conveyor belt on the other side and came down.

I followed this ham up one side and I realised that as it came down, it got the black mark on it. I shouted, 'Stop the conveyor belt and get me some steps!' I went up the steps and put my hand into the chute and there I found a rivet sticking up.

Now all I wanted was a hammer. I went to the engineer's store and got myself a copper hammer, which is the best thing to use on rivets. I went back and hit the rivet inside the chute and tested it for smoothness with my finger on it.

I asked for some hot water and soda as it had been a long time since it had been properly cleaned. We sterilised the conveyor, then we started the production again and that was it. What had happened was this: the rivet sticking up had collected a black mould and that was where the black mark had been coming from.

We both looked at each other and laughed with relief, then we went back to his office and he asked me what he owed me. I told him nothing – just a cup of tea.

He said, 'You're kidding,' but I was quite adamant that a cup of tea was fine. I went on my way but it had been a funny experience and I often think about it and have a chuckle to myself.

My own business was going along quite well and the girls were working hard; there was also a part-time man coming in and everything was running along nicely. When you use your supplies you know how much you need to order, and how often. We used a vacuum bag to pack bacon in, the air is withdrawn and freshness is maintained. You can estimate how many bags you use each week.

The local packaging representative used to call round and we used to buy bits and pieces off him; his name was Mr Roberts and he was a nice man. He used to sell good quality stuff.

One day one of the girls told me a gentleman wanted a word with me in the shop. He was a rural rep who told me he had got some vacuum bags and he could do me a special offer if I bought a large quantity.

I said I would have a look at them, but I warned him: 'Before you start, you won't be able to beat the price I get them for now, as we buy them in quantity and the quality is excellent.'

He was quite adamant he could beat the price. I asked him to show me his product. He went to his car boot and brought me the package and sure enough, the bags were beautifully presented and good quality and the price was cheaper than the ones I was currently buying. I placed an order with the new rep: the bags arrived and we started using them. After a short while, one of my girls told me we needed to order more bags. I could not believe this, as production had not increased. So I went down to the store and sure enough, there was only one box of bags left. This couldn't be right, so we opened the box and to our amazement, instead of a thousand bags per box as I had been told, there were only 750.

I thought to myself, 'That's a lesson, Maynard, no wonder they were so much cheaper! Put that down to learning!'

One day, a bacon producer from the north of England rang me. He was in very bad circumstances. When he cured his bacon, he'd get a green mould all over it, which he could not get rid of. They had scrubbed the factory from top to bottom thinking they must have had a bug in it. Evidently they had, but they had checked all the ingredients and still could not find the root of the problem. I was asked to come up and have a look.

I had gained the reputation, via reps and word-of-mouth, of putting faults right, and I agreed to go up north. We fixed a date for me to visit the factory.

It was a long journey to this northern town, and all the buildings were stone. I was directed by a local to the factory; it had a shop attached which was very nice. I introduced myself and was told I was expected. Mr Jenson, the manager, appeared and we shook hands and he took me into his office where he offered me a cup of tea. I refused as I wanted to get on with the job and see what was what.

He told me they produced the bacon and after a day out of either the salt or the brine, a green mould appeared on the bacon. If they wiped it off, it returned the next day. I was stumped for an answer but told him I wanted to go into the factory, starting with dry store and I checked all the ingredients to see if there was a problem there. If the saltpetre is damp, that will produce a green colour, or if you reduce the nitrate, that will also give you a green colour.

Mr Jenson told me they had already thought of that but did not

51

think that was the problem. I carefully inspected the seasoning room and it was nice and tidy, with everything well labelled. I put some of the saltpetre into a plastic bag and it was 100% dry, so I knew it wasn't that.

We then went into the factory and the maturing fridge and there the majority of the sides of bacon had indeed got a green tinge to them; that really shook me and I could not figure out what was wrong. So I got out all my test tubes and I tested the brines: the colour was fine, it was not the brines. What could it be?

I said, 'I know you are a busy man: could I just wander round the factory and leave it to me?'

He thought that was a good idea.

I started off with the layerage, and everything seemed to be fine there. Next was the slaughter house and that was all correct; very clean. I went into the cooling house where the pigs were taken after slaughtering, and they were all on spotless trolleys and then they went into the cutting room, and then into the curing room. I was really stumped and could not figure what the bacteria was or where it was coming from.

I said to one of the operatives, 'It's a funny system you've got here, putting the pigs onto the fore runners and running them into the cutting room.'

'Yes,' he agreed, but added that they found it a better system as they didn't have to swap them from one rail to another. 'It's easier on the trolleys, to take them into the cutting rooms.'

'So it's a new system you've got?' I asked.

'Yes,' he replied, 'We bought all the trolleys and the new system works well.'

I thought about this and asked if he would be kind enough to take all the sides of pork off and tip the trolley upside down. He looked puzzled, but I told him that was what I wanted him to do. He took the sides off the trolley and laid them on the bench, tipped the trolley over and there on the base of the trolley was the green mould everywhere.

I looked at the man and said, 'That is where your problem is.'

I went back to the office and told Mr Jenson the problem was solved. I asked him where he had bought the trolleys and he told me they were from a leather factory in Scotland, but he was adamant he had cleaned them all off.

I replied, 'You might have cleaned the tops, but you did not clean underneath and you have brought the bugs in from another factory. The best thing you can do is close the place for a day and sterilise the factory. I recommend you hire two steam cleaners, throw all the stock away and start again. Clean with common soda, a mix of bleach and water, and clean everything thoroughly. If you try and cut corners, you will not get rid of the mould.'

He said, 'This is going to be very painful but I will accept your advice.'

Two weeks later he rang me and said he had put my cheque in the post for my fee and he had done as instructed and everything was clean, no mould on anything, so fingers crossed everything had turned out alright. He said he had also sent me as a gift, an original *Meat Traders Journal* which was a hundred years old and he thought I would like this as a token of his appreciation and I still have that copy today.

I was working in the factory one day when one of the girls from the shop came in and told me a customer wanted a word with me. I went through to the shop and there was a young lady standing there.

She said, 'I bought some of your sausage and I have found this piece of glass in it.'

I asked her to come round into the factory and show me. She showed me the piece of glass, a large piece measuring approximately one and half inches by about three inches, and I asked her if this was the piece she had fetched out of the sausage.

When she said it was, I was totally amazed and told her we did not use glass in the factory at all; all the utensils were stainless steel and plastic. If it had been a piece of stainless steel we could have looked and seen if it had sheared off from one of the machines, but finding glass in one of our sausages did not seem possible. I asked her to leave the glass with me and to give her name and address to one of the girls and said I would be in touch with her after we had investigated.

She promptly piped up that she was entitled to compensation as we were selling a dangerous product. I assured her we would look into the matter, to discover how the glass had got itself into the product.

The next thing to do was to find any glass at all in the building: the nearest things were the cabinets in the shop, the only glass we had in the whole place. We checked the cabinets and there were no pieces missing. In fact her piece looked like a bit from a milk bottle. I had a thorough look through my spy glass and this confirmed it really did seem to be a piece from a milk bottle. We did not have milk delivered nor did we keep any milk bottles on the premises. The next thing to do was to go to the sausage machines and make sure this piece of glass could not have gone through the machine.

To make sausage, the meat is squeezed through narrow grids to make the fine texture and there was no way a piece of glass could have gone through this system without being ground up and stopping the machinery. So I knew it could not have gone through the mincer.

Next I moved my attention to the bowl chopper: this is the first process of sausage-making and if there had been a piece of glass in the mixture then it would have made a noise on the walls of the bowl. Also, at the chopper stage, hands are used in this process and I would have felt the glass. The last process is the sausage extruder, which puts the sausage meat into the skins and it has a fine-bore nozzle. I measured the piece of glass against the nozzle and there was no way it could have gone through this system, so I knew the glass could not have originated from the factory.

I rang the young lady up and asked her to come in. When she arrived I took her into my office, showed her the piece of glass she had brought me and said to her, 'That is the piece of glass you brought me. We have tried all the machines and it could not have come from our production line as it could not have possibly gone through our machines and stayed in one piece.' She was quite adamant it had come from the sausage.

I replied, 'I am sorry, my dear, I do not believe that. You or somebody else put that in the sausage to gain some reward out of it and the best thing we can do is to get in touch with the Public Health Authority and the police as this is a very serious matter as I am being blackmailed with a foreign object that did not come from here.'

Then her faced changed.

'I want to forget the whole business: it's been a mistake,' she said.

'Yes, it's been a mistake, a silly mistake,' I replied, 'But I am telling you now young lady, you want to go home and have a good think and put this wrong right and improve your standards because if you carry on like this you will have a very hard life.'

I was very cross and disappointed that people would do this kind of thing.

The local Government office for youth employment rang me and asked if I would take on a school-leaver to gain some experience. I thought this was a good idea as our industry needed the youngsters to become interested in curing so I agreed to participate.

The young man duly arrived for work one Monday morning. He looked presentable, seemed very polite and eager to learn and he told me his name was Richard but that everyone called him Dick.

I took him on a tour of the factory and told him there would always be work, as eating does not go out of fashion, and he laughed. I explained all about the industry and how our company worked, explaining that we were only a small company but we tried to produce good food and if he made the most of his time with us, it would be very rewarding. I really tried to impart as much information as possible.

I took him round the smoke house and showed him how it worked: I told him it was a bit like sailing a ship! And he laughed. I tried to make him feel at home. I told him that we would be getting the hams ready for the next week-end and that he could shadow me.

He seemed very interested so I thought I would show him how to present a ham for sale. I told him I would sharpen the knife for him and gave him the protective clothing to put on, including the protective gloves so that he could not cut himself. He wore a chain-link glove to protect his hands and a chain-link apron to protect himself. I told him he looked like a knight in armour and he seemed to think that was funny.

So we started off with the first lesson on how to trim a ham for display. I showed him how to keep the knives in hot water so that they would slice through the fat easily, and how to hold the knives correctly so as not to do any harm to himself. He seemed to be able to manage to make the ham presentable and I thought to myself: I might have found myself an apprentice here.

I enjoy teaching and sometimes you find someone who has a natural talent and I thought I might have found it in this young man. The time passed and when it was time to go home, he went out carrying a large hold-all. I thought to myself, 'He must have brought a lot of stuff with him.'

The next day he came and I put him to work with the women, packing the bacon. He seemed to enjoy that experience.

The end of the day arrived and he left carrying out his hold-all, which again seemed to have a lot of stuff in it.

I said to him, 'Excuse me: could you let me have a look in your hold-all?' as I thought I could see the shape of a ham in it. He became all flustered.

'No, no, my Father is picking me up, and I'm late.'

But I was insistent and when I looked in the hold-all, there were two hams in it.

I said, 'You are a very naughty boy. I was willing to teach you a trade and a skill which would have lasted you a lifetime and you could have earned a living for the rest of your life, but you have blotted your copy book on your second day here. Why did you do it?'

He looked at me.

'I was going to sell them.'

I was so disappointed.

'It will be no good for you here because now we could not trust you. You'll just have to leave now.'

To cut a long story short, I rang up the lady from the youth employment agency and told her what had happened. She came down and told me this was the third time he had done such a thing. I was a bit cross.

'You should have told me he had done that before,' I said, but she replied, 'No, we wanted him to have a clean slate because he promised he would not do it again.'

I told her we would not be taking on anyone else for work experience as it had put me off and I had not got the time to bother with it. I told her I had my own sorrows and at the minute I did not want anyone else's. We parted on good terms but I was sorry that one bad apple had destroyed it for anyone else.

CHAPTER 5
Salty Solutions and Frank Advice

In our industry, things jog along and you think everything is fine
and you are on top of the world and then something comes along
and puts you back to square one. On this occasion an incident did
put us back to square one: it was one of the most upsetting things
that can happen in the industry and I often think about it still. I can
laugh about it now, but I can assure you, at the time I could not
laugh about it. It really upset me tremendously but I have to admit
it was caused by one hundred percent stupidity.

It happened exactly like this: we had an order to produce 150
raw hams for a man who then cooked his own hams. They were
what you would call a 'fancy ham' and the difference between a
fancy ham and a dry salted ham is that the dry salted ham is done
with dry salt and the fancy ham is produced by using the brine
method. For the brine method, you inject the ham with an auto-
matic pump alongside the bone so that the meat cures from the
inside and outside at the same time, quickly, which is necessary if
you only want a very mild ham. Brine-injection is a well-practised
method, long established in the industry.

That particular morning we started production on the hams and
we had two injecting guns working: I had one and my co-worker
had the other. The hams are placed on a large stainless-steel table
and once the ham had been injected it was pushed from the table

into a large container with wheels. Once this container was filled, then it was pushed away and another container replaced it. It was a simple operation and took roughly two hours to produce all 150 hams for that order. The next stage was to drop them into the sweet pickle in the curing fridge. We checked all the temperatures and the density of the brine. Everything was correct.

Then we put everything away and battened down and I thought to myself, 'That's a good day's work done.'

My co-worker and I were finally cleaning the machines and the guns, when I picked up one of the two guns and I noticed there was a needle missing.

I said to my colleague, 'There's a needle missing!' and he replied, 'Yes, that's right, it came off.'

I could not believe I had heard him correctly. 'What do you mean, the needle came off?'

A silence followed.

'I lost it; it must be in one of the hams.'

I managed to say, 'You are kidding!'

But he wasn't.

I was trying very hard not to lose my temper. 'Why didn't you tell me?'

He looked at me. 'I didn't like to.'

'You do realise the position you have put us in,' I said. 'We have approximately two hundred hams in the brine with a foreign body in them, and we can't sell them.'

A long silence followed. I felt temper, frustration and anguish.

I said, 'That is utter stupidity.'

I thought the best thing to do was to go and sit down, have a cup of tea and think through the problem and how best way to tackle it. But I could not figure out how best to proceed. We had 200 hams (50 of our own) and one of them contained a needle, but we had no idea where the needle was. As it stood, we had lost the whole of the production. I had a reputation in the curing world as a problem solver but at that moment I was well and truly flummoxed.

A strong cup of tea, a calmer mind and half an hour later, I had an idea: the needle was metal, so if I could borrow a metal-detector, that should find it. When the hams were cured, we took them out of the brine and laid them all out in the factory, all 200 of them. A friend lent me his metal detector, and it took the best part of two hours, but eventually, we found the needle. In all the time I had been in the industry, I had never known such an episode of sheer stupidity.

I was working in the factory one day when one of my young ladies came to tell me I was wanted on the phone. A voice announced, 'This is John Cole speaking. I was hoping you could spare me a few minutes.'

He told me he had very salty bacon. That happens sometimes, but he said all the bacon was exceptionally salty and he could not find the cause. I advised him to check the brineometer, a glass instrument which checks the salt content in the brine because sometimes a fine crack can appear in the brineometer and this will alter the readings, so the producer puts more salt in than is necessary. He had not thought of that, so then I told him if that did not sort the problem out, to give me a ring back.

A week later, the phone rang.

'Hello, Maynard, John here, our problem has not gone away: the bacon is that salty you would need a gallon of water on the table when you had finished eating it!'

I replied, 'Well, that's no good. I can come down to see you in about two days' time; just give me the directions.'

Two days later I arrived at John's factory. It was quite a modern place, very clean, and I introduced myself to the receptionist who told me I was expected. He then asked me if I would mind moving my car.

When I asked him why, he said, 'That space belongs to one of our Directors who has recently died, and no-one parks in his space as a sign of respect.'

I said, 'I quite understand: we're all a bit eccentric in our trade!' and we both laughed.

I re-parked my car and collected my bag of tricks and went into the office. John had two sons in the business with him: one was sitting next to him and the other son was sitting at a desk. John went through with me how they produced the bacon. They explained they had a central store where they kept all the seasoning, described the densities they used for the curing, and all the timetables for the curing.

All sounded correct, yet when the bacon was ready, it came out so salty it was not edible.

'We've had to stop sales and there are about three tons of bacon which are not saleable. Is there anything we can do?' asked John.

I replied, 'Let's leave the salvage for the time being. The first thing to do is to find the fault.'

One of his sons piped up, 'Are you a clever arse at this job?'

I was a bit taken aback. 'No, I'm not a clever arse at the job: I've been asked to come here to help, but you have the rudeness of ignorance and you don't have the knowledge of the wise.'

Then moving on, I asked John if I could go into the factory to sort the problem out, if possible. I went into the factory and wandered about looking how their production-line ran.

I met the foreman and said to him, 'I've noticed the bacon you have in the storeroom is very red.'

He replied that it had been like that for a while but he had thought it was to do with the bacon equalising and the physi-ological changes during the curing and the temperature being too warm.

I then asked him if he would show me their seasoning room. I was led along the passageway to an enormous seasoning room, with hundreds of containers all marked with various seasonings. I asked him which contained the bacon and ham curing seasonings, and he indicated one.

I opened it, put my hand in the mixture and said, 'This is not seasoning: this is pure salt.' He looked at it.

'Is it?'

'Where's the salt?' I asked.

He pointed to another container and I lifted the lid up and had a good look.

'This is the curing mix: you've got the wrong lids on the containers, so instead of putting the bacon curing ingredients on, you have been putting salt on. And that is why the bacon is ten times saltier than it should have been.'

He looked at me.

'What a performance!' he said.

'Yes, well, it doesn't matter now: at least we have sorted it out and that is the main thing. The next thing is to make sure it does not happen again. Go and get me two glasses from the canteen, I want to check both containers, so I can ascertain my theory is right.'

He duly came back with the glasses; we put water into each glass and a spoonful from each container into the two separate glasses, and sure enough, it proved the wrong lids were on the ingredients.

I went back to the office, knocked on the door, and said, 'We have had a bit of luck: I think we have solved the problem.'

He was really pleased.

'That's wonderful,' he said.

I told him I thought the stock room should be re-organised to prevent a problem like this happening again: different-coloured containers, a better labelling system, so that the wrong lids could never again be put on the wrong containers, and the bacon ten times stronger than it should have been.

He asked me how we could salvage the salty bacon.

'I think the solution will be painful, but the problem can be solved,' I said.

I told him how to do it and he was really pleased.

He asked me how much he owed me, and I said 'Two hundred

pounds will be about right.'

He said he really thought the job was worth more than that, but I reiterated two hundred pounds would be fine. He wrote me the cheque and when I looked at it, I could see it was for three hundred pounds.

'You've paid me too much money,' I pointed out.

'No, I have paid you what you are worth and I may need you again in the future. I want us to end on good terms. My youngest son Jonathon was extremely rude to you and I want to apologise for him.'

But I told him it was not necessary to apologise. We shook hands and I felt pleased, because I could see he was a worried man and it had been a pleasure to be able to give him some help – it gave me great satisfaction. I headed off to the car park and there next to my car was the youngest son.

'I owe you an apology, don't I? I was foolish.'

We both shook hands.

'You're on the road to success when you've realised you have made a mistake,' I laughed and we parted friends as I headed off for home.

Life was trundling along nicely. I realised our product was right because we never had any complaints: and you can guarantee that if the product is not right, the complaints will come flooding in!

One day the phone rang and it was the BBC. They had done a national survey of many bacons and our bacon had won the *Best Bacon in England* award. They asked if we would go to London to collect the award, and I said it would be a pleasure: I felt very honoured with the accolade. I really did not want to go to London myself so I thought the best thing would be for Trisha and Anna to go to London to collect it, because after all it was a joint effort between us.

I had already made Trisha Managing Director because I thought it was the right thing to do. She was an excellent help in the business and dealt with many problems that I could not do. I thought this award was recognition for the many years of hard work and skill that she had put into the business, and our marriage as well, standing by me when times were hard.

Trisha and Anna duly went off to London to collect the award, bursting with pride, and I was aware they would fit in some time to do some shopping! So I stayed at home and looked after the bacon!

One day a huge lorry drew up in the yard. Emblazoned on its side was the logo of a major company which supplied all the large supermarkets with their bacon. The driver came into the shop and asked if he could buy some bacon for one of his customers. I looked at him.

'That's very strange. You represent a company which has hundreds of tons of bacon, so why do want my bacon?'

He could not quite meet my eye.

'It's for a special customer,' he mumbled.

'You've been sent here to buy my bacon so you can test it back in your factory,' I snorted.

He vehemently denied this but I told him I would not sell him any bacon and I was going to send him on his way unless he could tell me the truth.

'Well they won't be very pleased when I go back and tell them you won't sell me any bacon,' he replied.

I looked at him.

'You go back and tell them to send me an honest messenger next time and tell your bosses my Mother drowned all the foolish children at birth.'

True to form, a better messenger rang me a few days later. He explained he was the Production Director of this company and they wanted to come and see me. A few days later the Production Manager arrived in person and said he was interested in my product which they thought was unique. I told them I thought it was not unique, because all the recipes were traditional, but it was good.

He said, 'Can we talk?'

I told him I had not got a splendid office but that we usually had a cup of tea in the farmhouse kitchen, where the Aga made it cosy in the winter. I made a cup of tea, we sat down, and I said to him, 'Cards on the table: what do you want?'

He replied, 'Cards on the table, our bacon is unremarkable and we have been asked by our main customers to put a jazz in it. Then your name came up and that is what I am here for: can I buy some recipes off you so we can produce something different? If we send you twenty backs of pork, could you cure them in different ways?'

I replied, 'That's not a problem: tell me exactly what you are looking for.'

He replied, 'I want something people can say they actually enjoyed whereas at the moment, they just eat our bacon.'

I pointed out that it was difficult to achieve with mass production, because you have to put a bit of love into the special kind of bacon. It's rather like going to a restaurant that isn't top drawer but the food is fabulous: if you peep through the kitchen door and watch the chef, you'd know he put a bit of himself into it: all good chefs put a bit of love in it, and attention to detail.

He replied, 'Things are difficult in the industry at the moment.'

'They always are difficult in the industry,' I agreed.

'People want something that tastes delicious at a price that you can't produce it for, and everyone wants a bigger slice of the cake but sometimes the cake keeps getting smaller.'

He asked me which way I thought things were going and I told him I would need notice to answer that question and we both laughed. We agreed on the twenty backs and true to his word, they arrived. From this I produced about ten different varieties of bacon, and once they were ready, they were picked up. About a week later he rang back telling me the bacon was excellent and he told me he had selected three different flavours in particular. We arranged to meet the following week and for me to go down to his factory.

The following week I made the long journey there and eventually arrived at an ultra-modern factory with a commissionaire at the front door and two smart young ladies on the reception desk.

I told them who I was and that I had come to see the Managing Director; I was expected and they took me to his office. We shook hands and he told me to call him John. He introduced me to his Production Manager, and when we shook hands, I could tell he had been in the bacon industry a long time: it's a strange thing in our industry but people seem to have big hands. I have no explanation for this other than that we work all the time with our hands.

We commenced our tour of the factory, taking in the curing department, the sausage area, the slaughtering area and finally back to the office. John asked me if I had enjoyed my tour and I said I had.

He looked at me. 'That sounds a bit non-committal,' he ventured.

I replied, 'I am a guest in your factory; I would not comment.'

He pressed me again for my views and I came out with it:

'You must be very wealthy people to be wasting as much money as you are.'

He looked bemused.

'You think we waste money?'

'Oh yes,' I replied, 'But I am a guest here and it is not for me to say.'

He said, 'Well, now you've started you must finish.'

So I began: 'Well, the first problem is the toilets are at one end of the factory and when the female staff want to go to the toilet, they have to walk through two departments which takes ten minutes and by the time they've used the toilet and gone back to their station, that is half an hour wasted. With the amount of staff you have in this company, it would pay you to put some toilets up and down the factory for easier access.

'Also your fridges: the doors are too narrow so you can't get a stacking truck in them. This means everything has to be off-loaded and lifted manually into the fridge and then moved out the same way: the doors should be wide enough for the stacker trucks to go straight in and out.

'Your sausage area needs industrial curtains, by which I mean polythene curtains to retain the cold air: it would save you a lot of money.'

John was looking at me and nodding.

'It's just my observation,' I added, 'But another thing: the factory is backwards.'

He looked at me. 'What do you mean, backwards?'

'Production would work better if you had a flow line system: the pigs come in one end and the bacon out the other.'

He said, 'Well, I am amazed.'

He asked me if I would like to stop for lunch, which we had in the Directors dining room – very posh it was too! He introduced me to the other Directors and John told them I thought they were all wealthy people because they were wasting a lot of money. The questions came thick and fast, and I explained myself.

We enjoyed a very nice lunch, and they all took the criticism very well but, I thought maybe I should have not said anything. After lunch we went back to John's office, and he asked me to sort all the production and the new bacon lines we'd discussed. I agreed to sort out everything.

He then said, 'Will you sit down with me for five minutes? I would like to make you an offer. I would like you to be our Technical Director.'

I was astounded and could not answer straightaway.

'We would pay you very well,' he added.

'I have no doubt about that and it's a very kind offer but I am a

67

bit of a wild horse and I don't think I would take kindly to having a bridle these days. And I would have to have a bridle if I came here because sometimes I have to speak my mind and I know it doesn't always suit. You have made me a very kind offer but you can put everything right here yourselves as it's only common sense and you have a lot of common sense.'

We shook hands and I went on my way.

CHAPTER 6

Robbery, Bacon and Banquets

We had settled well into the new house. It was in an ideal position, well set off a main road with a long drive, shaded by a good many trees. There was a large car park in front of the house with a gravelled car park, so it was an easy access for anyone wanting to come down to the shop.

We had been in bed one particular night for about an hour when our daughter Anna, who was still working in the business, came to our bedroom door. She said she could hear someone outside. I told her old houses creak and make funny noises in the night but she was most insistent that she could hear someone in the drive so I got up and looked out of the window and lo and behold, it was a moonlit night and there were two men pushing a white van up the drive towards us. I knew then that they were up to no good.

I went downstairs and made sure all the doors were locked. The dogs were completely quiet which was unusual. I went back upstairs and opened the window to watch them. They were pushing the van down to the shop and I knew then we were going to be burgled.

I rang 999, asked for the police and said I thought we were being burgled so I was put straight through to the Duty Sergeant. I explained what was happening and the Sergeant told me not to turn the lights on, to try and keep the dogs quiet, and not to go outside under any circumstances. Meanwhile, I watched the men from upstairs. Out of the back of their van they took a bag and removed a tool from it, presumably to try and force one of the windows. There was a security light at the side of the house and as the men passed this, it lit up and I could clearly see them. I knew we were running out of time.

The next thing I heard was one of the downstairs windows being broken and there was still no sign of the police. I had my wife and daughter in the house and I knew I would be no match for three men. Once they had got into the house, we would be in trouble. I did the only thing possible: I opened the window and shouted, 'We know you are there and the police are on their way.'

Immediately two of the men ran from the side of the house, and with the third man jumped into the van. At first the van would not start, but it fired up on the second attempt and they took off with such great speed I thought they would run into the side of the house. Down the drive they went at a furious speed and were gone.

More or less immediately the police arrived, six of them standing there. They asked me what had happened and I told them they had missed the gang by about six seconds.

They said, 'We do have a procedure in which we can block all the roads, but we think as the attempt to burgle you has been foiled, they will probably park the van in a field until first light next morning.'

I said, 'Well, it looks as if the birds have flown,' and the policemen agreed with me.

The boss told me he would get all his officers to search round the premises and the fields to make sure they had gone. They a good search of all the outbuildings and fields and came back to tell me they could not find any trace of the gang. I was asked if I could give a description of the van, and I said it was a Ford Transit, with the logo of an arrow on the side, which I could see because the security lights were on. So the police put out an alert on the van.

I asked the police if they would like a cup of tea or a drop of whisky – in fact, I said I felt I needed a whisky and they agreed with me. Just then, one of the policemen who had been searching the grounds came in with the hold-all bag the men had left behind, next to the window they had smashed.

It contained the tools of their trade, and the police told me not touch the tools as they might be able to lift some fingerprints from them. In the bag I could see a frightening array: a couple of hammers, screwdrivers, and a long iron bar.

The policeman said, 'God must have smiled on you this evening. These were real professionals and God knows what they would have done if they had got into the house, because you are well off the road and nobody would hear anything. Tomorrow I suggest we send our Crime Prevention Officer along to give you some security advice as you have been very lucky tonight, exceptionally lucky.'

It was a moment of truth. I knew I had been close to a very serious incident and if it had not have been for Anna hearing the gang pushing the vehicle up the drive, it could have been a very different story.

In due course the security officer came, a very pleasant man who walked round the house, and said, 'It needs a lot to give you some security, because it's way off the road. All you can do is make the house more difficult to get into. If they are determined to get in, you can't stop them, but you can make it more difficult for them to get in. All the windows will need double locks and some of the windows need bars on.'

I said, 'It will be like a prison,' but he replied, 'Unfortunately today, that is how it is: if you want some security for your family, you have to make sure it is more difficult for them to get in.'

So we did as he advised: we had locks put on the windows, but when he told us in some cases thieves take the windows out, we had the local blacksmith to put bars on all the windows downstairs. I was a little bit disappointed because this was my home and I was making it into a fortress and I thought to myself, 'In this country, we have lost more than we have gained – what a society!'

The phone rang one day and one of the girls came to me with a message. A Mr Strange was on the phone for me. He asked if he could come and see me in person as he had an unusual request and he thought I was the man for it. We arranged a day and a time for him to come and see me. When we met, he was in his early fifties and was the type of man to call a spade a spade.

I said we would go up to the kitchen and have a cup of tea, while he put his unusual request to me. Once we were settled comfortably, he told me he was an agent for a member of the aristocracy who had decided to have a medieval banquet but it had to be authentic. My name had come up and I had been recommended to produce the meat side of the banquet.

I asked him exactly what he wanted, and he replied: 'Ten suckling pigs, five or six Old English hams, a hundredweight of medieval-style mincemeat; and then we want the original sausage.'

I said, 'To be very honest, Sir, I can do it, but do you realise the expense of it?'

He said he did. I told him the pigs would have to be wild boars to make it authentic and the hams would have to cured specially. The mincemeat would have to be made up from the original rare breed cattle of the day, and the sausage would have to cased in beef runners (intestines) not the later pork runners. Medieval sausages were thick and they were boiled, not fried, so they needed a thick skin to withstand the boiling water. All the meat would have to be seasoned with medieval herbs and spices, which was going to take time, and time is money.

He replied, 'Money is no object: this is a one-off and we want it right.'

So I started work. I decided to make the mincemeat first, because it improves if made a month earlier, and left to mature. To do the mincemeat correctly I needed to find someone who made earthenware jars as this was the traditional receptacle for medieval mincemeat. I found a firm who made these jars. I made the authentic mincemeat, put it in the earthenware jars and sealed it with a canvas lid. We wanted to be authentic right to the last detail, so we even contacted the apple research centre to find which apples would have been used in this era.

Fortunately the recipe book my old Apprentice master, Thea, had given me recorded these recipes but they were out of favour because they were too salty. I was pleased I had kept them as they had come in handy on this occasion.

The next stage was to cure the hams. Now medieval hams were cured entirely differently to today's hams and again, they were very salty as they had to keep for a long period of time. Our tolerance of salt today is not very high, because so much of our food is processed, with a high salt content. I contacted an Essex firm that still produced pure English salt, a quality product with a large grain, excellent for curing. Droitwich also produced this medieval salt. We acquired some traditional Tamworth pigs, our native breed. We applied the special salt and turned them every day for three weeks. Then we washed them down with vinegar and air dried them, slowly

We also discovered mutton hams were a popular medieval dish, as there was an abundance of sheep when the English wool trade was at its peak.

So we were proceeding steadily with the job: the mincemeat was done; the hams were curing nicely. When the hams had matured, we sewed them into cotton bags – although we later realised we should have used linen, as this was the predecessor to cotton!

The next thing to do was the suckling pigs, and this was a very interesting process. We discovered that, traditionally, the suckling pigs were painted on the inside with brandy and then covered with pork fat. Then the stuffing was made, put inside and stitched up, the outside was painted with butter and salt and the pig would then roast nicely on the spit. We made a wooden block to put in the pig's mouth so that after they had been roasted, you could put an apple in for authentic decoration. We had done our homework and we were slowly getting there.

Next was the sausage. The sausages in those days were made without machinery so we cut all the meat by hand, crushed the salt, added the parsley and sage according to the original recipe, then we put them into the large beef casings and made them into sausage rings. We tested them and were satisfied: we were well on the way to completion.

In the meantime, we agreed with the agent we would also cure a mutton ham. We knew someone who bred the Herdwick sheep, so we acquired one of his older sheep off him and we cured the mutton ham. This had not been done for a very long time, and we all felt very pleased with ourselves.

Duly the day came and the meat was all collected and we thought it was as near-authentic as we could possibly get it.

About a fortnight later, Trisha told me there was a personal telephone call for me in the house.

The agent was on the phone and he said, 'We had the traditional banquet and it was one hundred percent successful and I wanted to ring you to thank you personally.'

I told him I was just doing my job and he replied, 'No, you were not just doing your job; you are a craftsman and we are very grateful to you. Now I have a gentleman here who would like a word with you.'

And the gentleman who had commissioned the banquet thanked me very much. The night had been a success and that was payment enough for me – job well done.

Here is a medieval recipe and even if you don't make it, some of you may get a bit of enjoyment reading about it.

Mutton Ham

Ingredients
2oz thyme, 3oz marjoram, 6 bay leaves, 2oz black pepper, 3lbs bay salt (large granules), 3 gallons water, one large mutton leg (24 hours before use, cover in salt to cleanse), half pint strong vinegar. Cayenne pepper (to dust).

Method
Put water in container, then add herbs, salt, pepper and boil. This produces the brine. Wash dry salt off mutton leg. Place in brine for 10 days. Remove from brine, wash off and dry. When dry, paint wth vinegar, then rub surface with cayenne paper. Thoroughly dry again. This is now ready for smoking.

Smoke
Smoke in 2 parts oak, 1 part peat, until golden-brown. To eat, cut very thinly.

This was a delicacy in the Middle Ages, particularly eaten in Wales, Scotland and the North of England.

I used to start quite early in the morning in my little factory: it gave me time to gather my thoughts and plan the day. I liked to start about half past six; none of the girls started until later, and it gave me time on my own.

On one particular occasion I was working away early when there was a knock on the double doors of the factory.

There stood a young lady who said: 'Are you Maynard?'

I told her I was and she asked me if I would teach her to make bacon. I asked her why she wanted to know and she explained that she wanted to earn some money, she felt that everyone wanted good bacon but couldn't get it.

I was rather intrigued. It was only quarter to seven in the morning and here was a young lady on a bike, wanting to make bacon.

I said to her, 'You realise it is a very committed job and you need a little bit of equipment to do it. Can you tell me your circumstances?'

'I am a young mother with three children who've started school: I've got time in the day to do it and I live on a council estate, where everybody says they would like some nice bacon. I think I've got the aptitude to work in the food industry.'

I was a little bit taken aback. I had had some requests in my life but nothing like this at this time of the morning! I asked her what her husband thought of the idea and she told me he had left her and the three children a year ago and she had to make her own way in the world, using her free time in the day.

Sometimes in life, somebody wants assistance and you know you have to give it. It was inconvenient and I was busy but I always thought if you want a job doing: ask a busy person. It was a commitment on my part but I told her if she wanted to learn I would teach her.

So I told her I started at half past six in the mornings, and if she wanted to be there so many mornings a week, I would teach her. She agreed. As she walked out of the factory and rode off on her bike along the drive, I thought to myself, 'That is the last time you shall see that young lady.'

How wrong can you be?

When I went down the factory at half past six the next morning, there was a bike leaning up against the wall, and there she was. I said, 'Good morning. I did not expect to see you,' and she laughed.

'Come in and I will find you some protective clothing, but first of all, tell me your name,' I said.

She told me her name was Ella, which I thought was a nice name. I went down to the storeroom and found her the necessary things: she was so tiny she looked swamped, but I started to teach her and found she had an aptitude for bacon curing: she was a joy to teach and she took to it like a duck to water. Some of the questions she asked sometimes had me scratching my head.

I had a lot of pleasure from teaching her, and in a couple of months the time had come for her to branch out on her own. I asked her where she was planning on producing her bacon.

She said, 'I've painted the garage out and I am going to do it in there.'

I bit back a smile.

'In the garage?'

She nodded her head at me.

'Well, to start with you will need a dustbin, a hundredweight of salt, some saltpetre and some sugar.'

I told her where to buy her raw materials to start her off. I told her she would also need a small slicer and some good quality greaseproof paper, the best available – if you are producing a good product you want to wrap it up in a good coat! And she laughed. We both thought and acted in the same way. Then I asked her how she was going to deliver the bacon and she replied, 'On my bike.' She explained she had some potential customers on her estate who all said they would try it. I thought it was a wonderful idea.

So she bought herself a dustbin and I gave her some salt, sugar, saltpetre and my old machine, a gift worth giving because everyone needs a start in life and if you plan to sow happiness, it's a good crop to reap. And that is how she started.

Ella came to see me sometimes to tell me how she was getting along and to pick my brains. She arrived one day and told me some of the other mothers were helping her. I asked her to bring me some of her bacon to taste and, true to her word, she brought me some and it was very good.

People think bacon-curing is male-dominated but they would be wrong because the best curers I have ever come across are ladies. In America, the Quaker lady I met when I was young was excellent. I often wonder why and I think it is because ladies pay attention to detail. So if there are any ladies reading this book who feel they might have the gift for bacon-curing, remember: opportunity knocks very softly.

One day Ella came to see me, very excited. She had progressed in the world and now had a small van.

She said to me, 'On our estate is a butcher's shop and the butcher is going to retire. I think I would like to take the shop over but I don't know a lot about the fresh meat side of the industry: would you help me?'

I told her I would give her any advice I could: she took the butcher's shop over and she made an excellent job. She also had a delivery van on the road and she employed about seven people, going from strength to strength. She was an excellent curer and an excellent business lady, but her judgement in men was terrible. She finally became involved with a new man, they married and within twelve months the business collapsed and they ended up with nothing. That was very, very sad and I was very upset for her about that.

Looking back over that time, I remember one or two of my customers in particular: some people stand out in your mind. I became friendly with a very nice couple. One day the husband came into the shop and we got talking. He said he wanted to do something special for his wife as it was coming up to her fortieth birthday and he wanted a special gift for her. They had both seen a lovely brooch in one of the jewellers but it was very expensive so he thought he might sell a few shares to buy her this and I told him

I thought it would be a lovely gesture because we only live once and happiness is a commodity we should seek.

Apparently he bought the brooch and, being a bit of a romantic, he thought on the day of her birthday he would get up early and leave the brooch in a box of Black Magic chocolates. So he left the box of chocolates by the side of the bed before he went off to work and she in due course saw the box of chocolates and was very disappointed with it because she thought a fortieth birthday was worth more than a box of chocolates. Later that morning her cleaner arrived and the lady of the house decided to give the chocolates to the cleaner.

When her husband arrived home that evening he asked her if she had liked her present and she told him she thought she was worth more than a box of chocolates on her fortieth birthday, and that she had given it to the cleaner in disgust.

A look of horror crossed his face and he said to her, 'I put a diamond brooch in that box of chocolates for you.'

His wife quickly put her coat on and ran down to the village to the cleaner's house, explaining the mistake.

But the cleaner refused to return the brooch, saying, 'No, you gave it to me and I am keeping it.'

It was a small village but the cleaning lady wore it on her coat whenever possible. The marriage did not last very long after that: what a tragic tale!

We used to keep chickens for our own use and we sold the surplus in the shop. They were rare breed chickens and laid a brown spotted egg. On one occasion I was going to collect the eggs in the morning when I opened the chicken house door and everything was dead. The chickens had been gutted, their throats torn out and it looked like carnage. I had seen what the foxes could do but I could not figure this out as the fox could not have got into the chicken house.

I had never in my life seen anything like it: every chicken had been torn apart. There were heads, legs, bodies everywhere and I could not work out what could have done this terrible thing. There was nothing else for it but to set to, clear up all the mess and bury what was left of the chickens. The sight really sickened and upset all of us. What kind of creature had killed thirty chickens just like that? I knew from previous experience it was not a fox.

We cleared everything up and bought some more chickens as we liked to have fresh eggs. About a week after the new chickens had arrived, I went to the chicken house one morning and the same thing had happened again: absolute carnage! For once in my life I was stumped; I could not figure it out. I had lived in the country most of my life and I could not imagine what kind of animal was killing the chickens.

I was walking the dogs one morning across the fields, a time I always enjoyed as it gave me time to reflect and I liked my own company with the dogs. As we returned from our walk and entered the farmyard the dogs were having a sniff round when there was a terrible cry. I had a dog that would make a firm stand at most things and I had never seen him back-peddle as he did that morning.

As I went over I could hear a hissing noise. I went to investigate and there was a mink, a vicious mink, and with no more ado she attacked the dog! It was a very frightening experience. Fortunately I always carry a stick with me so I went for it, gave it a powerful blow and it shot away.

I looked at the terrified dog which had lost a good piece of its ear. I thought to myself: what a dangerous animal to have in the countryside! I made some enquiries and found out that someone down south had let lots of these animals out into the wild and I thought what a very irresponsible thing to do. I was not the only one who was having trouble with them and I never forgot that and nor did the dogs. When we went for a walk again the dogs would give that corner of the farmyard a wide berth and I can't say I blame them. It had shaken me to my very core.

In our industry you have to prepare and 'set out' before you open for business each day so you start work a couple of hours before the shop opens. It was a long job for us to set out as we did roughly thirty different types of bacon, in about three hundred individual packets. The display cabinet was large: about 28 foot long by about two feet wide, which was emptied out and cleaned each night. Not only did the bacon go in each morning, but various hams and sausages too, so it was quite an operation.

I used to start work about 6.30 in the morning and wheel all the products out of the fridge. I liked to go down early on my own so I could take my time and check every packet.

I was very proud to produce so many different varieties. I checked the packets carefully because sometimes you would get a 'blower'. This is a packet with a faulty seal. Our bacon was vacuum packed to stop the bacon oxidising; if you leave bacon out in the air, it loses it colour, which is one of the reasons why bacon is vacuum packed.

Early one morning I was working away steadily when I heard a vehicle pull up outside, which I thought unusual because customers don't usually arrive until about half past nine. The van door opened and two men got out, followed by seven children.

The next minute some of them burst into the shop while at the same time I heard someone in the factory. I went to investigate but as I went into the factory, the children darted behind the counters and I was totally overwhelmed. It was only seven o'clock in the morning!

I ushered the children back from behind the counters, but the men who were with them did not seem to be bothered at all. Then I saw the children helping themselves to the bacon in the cabinets and also to the butter and cheeses. They were stuffing their pockets, and every time I tried to stop one, another would help themselves.

I was totally marooned in the shop – I had set the till out so I did not want to leave the shop to get help. I couldn't do anything at all. When they had quite finished filling their pockets, the men bundled them back into the van and drove away. I was so taken aback. It had all been so quick: they could not have been in the shop more than five minutes.

I did a quick look at the stock and reckoned they had run off with a couple of hundred pounds-worth of bacon, cheese, butter and eggs. I went back up to the house, dialled 999 and asked for the police.

When I got through I said to the policeman, 'You are not going to believe this,' and he replied, 'Listen we've heard one or two good stories, but tell us yours.'

So I told him what had happened.

'I've just had seven gypsy children and two travelling men come into my shop and they've cleared me out of about two hundred pounds-worth of stuff. I wondered if you can do anything about it.'

I heard the policeman chuckle.

'By the time we catch up with them, the bacon will be gone and they will be well on their way to another part of England.'

I thought to myself, 'What a situation: what a country!'

In all the experiences in life I have had, I have never felt so helpless or alone, even though my wife was in the house.

I decided that as this lot of birds had flown, in future I would keep the gates on the drive locked until the girls started arriving for work.

I thought to myself, 'How things have changed - and not for the better!'

CHAPTER 7

Auctions, Weddings and Happiness

Sometimes in life you can have a bit of good luck: you're in the right place at the right time. Not many times in my life has that ever happened, but on this occasion it did. I had always liked beautiful objects and I enjoyed going to local auctions. It was a change from my normal working life and it's good to have another horizon to look at, where you meet different people who have different outlooks. I enjoyed watching the auctions and sometimes I would buy something.

On this particular occasion I had gone to a general sale and there were some nice pieces there, but nothing I really wanted. It was an afternoon out and I would enjoy just sitting there and sometimes I would play a game of looking at the different lots and imaging how much they would fetch; whether by luck or judgement I would be right on some of the prices. On this occasion the auctioneer was not doing too well: people were moving about, and several of the lots weren't reaching the reserve price and the auctioneer was getting a bit grumpy!

A lot came up of a well-made box of cutlery which the auctioneer said belonged to a good home, somebody who had plenty of time on their hands to clean it, as the cutlery was absolutely black! He asked for bids and when no-one responded, he said £60: still no response; £50 no response; £40 still no-one answered.

The auctioneer was fast losing patience:

'Will anyone give me £20?'

I put my hand up!

'I'll give you £20,' I said.

The auctioneer looked round the room and asked for any more bids – silence – so he said, 'Going, going, gone,' and banged the hammer down. I had bought myself a box of cutlery.

I went up to the desk and paid the £20, then the auctioneer gave me the tray of cutlery and told me there were a further nine trays!

I said, 'You're kidding!' but he confirmed that the lot had a total of nine trays of cutlery.

So I backed the truck up to the loading bay and put all these trays of cutlery in the truck. Every one of them was completely black and tarnished.

By now I was interested in all these trays of cutlery so I went back to the auctioneer and asked him where it had all come from and he said, 'We cleared out the crypt of an old church, and we did not know whether to leave them or bring them, as they were in such bad condition, but as you only paid £20 for them, it doesn't matter.'

So off home I went with all these trays of cutlery and I left them in my workshop for the night.

The next morning I went down and had a proper look at them. I rubbed one of the spoons and on the back of it emerged the words Mappin and Webb, so I then took one of the forks and one of the spoons up to the house and ran them under the hot tap to clean some of the muck off and there on the spoon and fork were also the words Mappin and Webb and the silver hall-mark.

I realised then they might be solid silver.

I rushed into the office for my magnifying glass and had another look. Sure enough, there was the silver mark. I couldn't believe it.

I went back to the cutlery trays and took some more of them to the house and washed them and, yes, the mark was on them all. I could not believe my luck. I had bought approximately four hundred knives, forks and spoons, all matching, and they were very old, as the forks only had three prongs.

From then on, after I finished work every evening, I would go and clean the cutlery and polish it until it gleamed. I remembered when I was a little boy how my Grandmother would clean her silver: she was a very careful lady or, for the want of a better word, a very mean lady and she never bought anything.

To clean her silver she used to put a couple of tablespoons of soda in an aluminium saucepan with a quart of water in it, put the silver in the pan and let it slowly come to the boil, remove the cutlery, rinse it in boiling water and dry it. I adapted this by lining a bowl with tin-foil, adding the soda and cold water and that did the trick. My Grandmother always wore a pair of cotton gloves to handle the silver once it was clean, so as not to mark it, and again I followed her example.

It was exciting to watch the silver coming out sparkling and clean. I was so thrilled I rushed up to the house to show Trisha. 'God's smiled on us: it's solid silver and look at the condition!'

We looked at each other and we both agreed: 'There will be turkey for Christmas!'

I went down to the workshop after each working day and carried on cleaning all the silver. I would put them into tens as I cleaned them: ten knives, ten forks, ten spoons. I put them back into the wooden boxes, lining them with cotton and it took me the best part of three weeks to finish cleaning them all.

I asked Trisha what we should do with them and she felt we would never use them but that we could do with the money, and we should sell them. I knew an antique dealer in the town, so I rang him and asked if he knew of anyone who dealt in silverware. He put me in touch with a gentleman who worked in the London area, where most of the cutlery is sold because that is where there are a lot of restaurants.

I rang the dealer and told him I had some cutlery for sale. He wanted to know how much I had got and who the maker was, and when I told him the maker was Mappin & Webb, the phone went quiet.

He said, 'You've really got that much?' and I confirmed I had. He asked me if they could come and see it and the day duly arrived. We had laid all the clean cutlery out in the dining room and it looked a bit like Aladdin's cave. He and his colleague sat down at the table, picked some of the cutlery up, looked at it with his jeweller's eye glass and he said, 'What are you asking for all this?'

I said I had no idea how much it was worth.

He looked at me and said, 'What I will do, if you are interested, is make you just one offer for it: I won't go up or down, I will give you three thousand pounds for the whole lot.'

I shook his hand before he changed his mind, and true to his word he went with his colleague to the car and came back with a very smart case, opened it and paid me in fifty pound notes, the whole three thousand pounds: I hadn't seen so much cash for a long time. We loaded all the silver into his vehicle, he shook me by the hand and told me if I ever came across any more silver, to give him a ring.

He said, 'I think we've both done alright.'

I smiled at him.

'I've no regrets and I don't think you have,' I said, and we left on good terms.

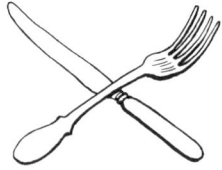

Time waits for no man and in our family, our daughter Sara had decided to get married, so it was all hands to the pump to provide a nice wedding for her. I said I would help organise the wedding and was told quite bluntly that it was a ladies occasion and all the men were required to do was to turn up on the day! And you know that is very true and on reflection I thought that was a good thing.

If there are any foolish men out there who want to interfere in a wedding, here is a piece of good advice: don't interfere!

So without further ado, we were on the way to organising a wedding, with me in a minor role and the women in the major role. We contacted an outside caterer to quote for the food and a smart man with a clip board and a sharp pencil duly arrived. He sat us all down in the dining room around the table and asked us what we required.

I kept a very low profile, while Trisha told him there would be approximately one hundred and fifty guests, and that we would like a good wedding. He did some calculations: we would need a marquee, spirits and beer, something to toast the bride and groom. We were all very quiet sitting around the table.

He did some more calculations, looked at us, and said, 'I think I can do it for nine thousand pounds.'

There was silence. Trisha found her voice and asked him to send the details in writing, and we would let him know. But Trisha already knew we could not afford this figure, as it was a tremendous amount of money. After the gentlemen had gone we all sat round the table, gathered our breath and decided we could do the wedding ourselves if we all took a part in it. Sara would organise the marquee and all the tables, Trisha would organise the waitresses and my role was to produce the food. We were all confident we could do it.

The date was arranged for the wedding, in the pretty, local church. I sat down and thought about my role: it was a special occasion and I wanted it to be right so I made my mind up that I was going to put my best effort in.

One of the first things I did was to contact the wine merchant to organise the wine and when you realise how much wine you want, it's staggering! We reckoned we wanted three hundred bottles, which worked out roughly four bottles per table, six people to a table. It seemed a tremendous amount to me but I was assured that that was correct for weddings, with the toasts and everything.

We went to the wine merchant's, quite a large place, and told him our requirements and the date. The wine merchant suggested we would need the wine delivered three or four days prior to the wedding as it would need to settle, and I took his advice because I always think it is wise to listen to people who are experts in their own trade. He suggested we put a bottle of red and a bottle of white wine on each table to start, and we ordered three hundred bottles of wine altogether, plus a small barrel of beer and an abundance of orangeade, cider, lemonade, bottled water and it felt like we were equipping an army out to go to fight the wars.

He said he could supply the tablecloths, cutlery, all the glasses and he said the contract was that if there were any damages, there would be an extra charge. He asked us which wine we would like: we walked down all the aisles of wine looking at the different varieties and I spotted a man I knew, also browsing.

He asked me how I was, and I replied, 'We are in a very painful position at the moment: we are buying wine for a wedding,' and we both chuckled.

We chatted and he asked me if I would like a bit of advice on the wine and I said I would. 'If you have got a moment, I will take you along these aisles and show you which of these wines is good value for the price.'

He pointed out the less expensive red and white wines which would have just as nice a taste as the more expensive ones, and we decided on these wines and I thought it was kind of him to give me the advice. The wine was delivered a few days before the wedding so the wine could settle down: there is more to weddings than you realise, and if you listen you can learn.

Now was the time to organise the marquee. We rang a few firms up and three quotes came in at different prices, ranging from seven hundred pounds to fifteen hundred pounds. Naturally we had the cheapest, and I think it was the nicest one, with a lining, and covers for the tables, so we were well on the way to success. The marquee came with the tables and chairs, so everything was now organised.

Next I had to organise the food, so that task began. We decided on a cold buffet, as a hot meal would be difficult for 150 guests. We wanted ox tongues, and I thought I would prepare four different flavours: a sweet tongue, a garlic tongue, a smoked tongue and a spicy one. Then I decided I would do roast beef, so I thought a plain roast beef, a cured one, a smoked and a spiced beef, so that gave us four types of beef.

Everyone likes the hams, but the roast ham would have to be cured especially, because a roast ham should have a delicate flavour: today where the mistake is made in the industry is that they take a ham and roast it. But hams for roasting and boiling should be cured in a special, gentle curing mix to retain as much of the flavour of the original ham as possible. And that is what I intended to do. So basically we had a nice menu now: we had the tongues, the beef, I was going to produce two types of ham, one roast and one smoked, not a heavy smoke; a gentle smoke.

Next were the sausages; then I decided I would also make a salami. I remember working years ago in the States with a man named Mauro who was a wonderful salami maker. In America all the meat departments are clearly defined, so a sausage maker is a sausage maker and a bacon curer is a bacon curer and the manufacturers stick to this.

This man Mauro was a star and he made a wonderful salami. As his name suggests, he originated from northern Italy and I was lucky enough to learn from him his original salami and that is what I planned to do: an original, special type of salami. I was quite happy with all the things I was producing for the wedding; I just wanted it to be special as I would not get another chance to do this for my daughter Sara.

I wanted a few more items so I could cater for everyone's taste, so I thought I would produce a luncheon meat and then I decided we would also have a nice pâté. In the end I wanted to do two types of pâté: a smoked pâté and a liver pâté. Liver pâté is best made using pigs liver in my opinion, and it's also important to use a certain type of bacon in it: streaky bacon, which adds to the flavour. So we had the two types of pâté, the smoked and the plain, and the menu was slowly beginning to build up.

The next thing to do was the pies; I decided it was no use making small pies so I decided to make the Grosvenor pie, which is a long pie made with pork with hard-boiled eggs down the middle. I made the special mix with pork, my own flavourings and I used the eggs from our own chickens.

I thought we should also have a pork loin, not a leg but the loin, and I decided I would remove the bone from the loin. Attached to the loin is the streaky bacon so I decided to leave this. This is what is known as a middle of pork. Once boned out, I used a meat hammer to flatten it, then I made a mix of coriander, butter, eggs, breadcrumbs and sage and spread this over the loin, then rolled the meat up, tying it with string. I scored the rind, then rubbed it with butter and salt, and baked this in the oven until it was all golden brown. I was quite happy with the menu. I had put a lot of love and years experience into this menu, to make it something special. All these recipes, the methods, the ingredients, I intend to pass on to the future generation of curers.

We called the wedding 'D Day'; I suppose it was an apt name. The next stage was for me to have a dress suit, which I objected to. I wanted to wear one of my own suits, but I was over-ruled.

The whole thing was like arranging a battle. The women had

taken charge and were in full flood and no matter what anyone said, they knew what they were doing: so any sensible person stood back and I hoped to be one of those – I let it flow and did not interfere!

Four days before the wedding, the marquee men arrived and started to erect the marquee like true craftsmen – within two hours it was up, they had laid the carpet and the electrician had installed the lights. It was time to start laying all the tables, the ladies had arranged a seating plan; it was all done very well.

The day before the wedding, we had to get up at two o'clock in the morning, to go to the Birmingham flower market, where we filled the truck with flowers for very little money. Now everything was in place: the marquee was ready, we had arranged the flowers in church (they had even made their own headdresses), the dresses were ready, the bouquets: it was a self-styled wedding and we'd pulled it off. I was very proud of the achievements of the ladies.

The morning of the wedding arrived, bringing with it a little bit of panic. A hairdresser was on-hand to style all the heads of hair. I kept a low profile and dressed up in the suit which I did not one hundred percent like and did not want to wear.

I decided I would change and get back into something more comfortable at the first opportunity, but for the moment I played my part, and all went to plan. The photographs were taken in the house, Mother and four daughters. In a moment of pride I looked round at the scene and it had been a long journey but it was a moment of great happiness and a new beginning for my daughter.

The cars arrived and we went to the church, a lovely thirteenth century building in a fine setting, and I walked up the aisle feeling as proud as punch. The wedding service was soon over, the photographs were taken and we went back to the marquee to celebrate.

We had arranged to have a classical quartet to play and I highly recommend this to anyone arranging a wedding: it was a lovely touch as we dined. After the speeches, celebrations continued and unbeknownst to me, the family had arranged for a jazz band to play in the evening. They arrived and set up their instruments, more evening guests arrived and the buffet was restocked for them. There was a free bar and I have never seen anything like it in my life. There were drinks everywhere! My daughter had left at one o'clock in the morning so we decided, as it had been a long day, we would make an announcement to say goodnight.

As the guests filtered away into the night, saying their good-byes, we noticed all the half-filled glasses of beer, wine and spirits left untouched. I found this very sad. We all started to clear up, and as I went out of the marquee I nearly tripped over one of the guests lying in a drunken stupor, wrapped round one of the marquee posts. I asked for assistance; we took him back into the marquee and laid him on his side, then went round checking to see if there were any more casualties and we found three more people passed out, whom we also took into the marquee and laid on their sides.

The drummer of the band which had been employed to play in the evening, had drunk so much, he was riveted to his seat behind his drums with his drumsticks in his hands, unable to move.

I asked the manager if this was usual and he replied, 'It is your own fault for supplying so much free alcohol!'

So that was the end of the wedding, a happy day that kept us laughing for a long while afterwards.

One day we had a phone call from BBC radio's farming section, from a programme called *On Your Farm*. They said they had heard we produced farm bacon from just about every region of the country. I told them we produced about thirty different regional bacons, plus about four different types of smoked bacon. They then asked if they could do a programme on us, and I agreed.

They duly arrived, set up all their recording equipment, and a very interesting lady called Henrietta Green came with them. It was the first time I had met her: she was a very professional lady who knew everything worth knowing about food. She asked some very good questions and I really enjoyed doing the programme with her. She wanted to know where my pigs had come from, how we looked after them, the feed we used, how the curing was done, what different kinds of regional bacons existed. I also told her about the sausages we produced with no additives, just meat and seasonings.

There was no spin with her. She knew her food and she was a professional interviewer. She asked my views on mass-production of meat and I told her about the industry joke that butchers never eat their own sausage because they know what's in them and they never eat anyone else's because they don't know what's in them! And she laughed.

I felt very pleased that I could enlighten people on the radio about the farming industry, especially publicising the farm shops selling bacon. And I said I thought farmers would be better selling in their own markets, as they were at that point having a very difficult time of it in the supermarkets.

Henrietta asked me lots of questions about curing and I told her every time you cured, it was a challenge, as you could never be sure if you had got it right until the end, as so many things play a part in the final product: the weather, the temperature, the original pig. But if you work on the premise that you know where the pig has come from, you know what they have been fed on, and you know the method of slaughtering, it all plays a part. If you get the pigs excited prior to slaughter, the cure never takes, so it is very important that you know where they come from.

We discussed regional bacons and Henrietta wanted to know, for example, why the Staffordshire Black was a black bacon. I explained it was the treacle made it black and the honey made it sweet, and that was how many people like their bacon in the Black Country. She asked why the Welsh bacon is so salty and I explained that the Welsh just liked it that way. It's all a question of taste: Cheshire bacon is similar to Wiltshire, a sweet kind of bacon;

in Devon they cure with cider which gives a fruity note. So every county in England cures different kinds of bacon.

As well as curing over thirty different varieties of bacon, we used over twenty different types of sugars and that surprised Henrietta, especially when I told her there were also about fifteen different types of salt. The secret in curing is basically the proportion of the ingredients, and the flavour of the different salts, the different sugars, so they form a unique base for your cure.

Every time you cure, it is a challenge and you have to have your wits about you, you need a bit of skill and a lot of love, well mixed together, so the product comes out right. And it has got to be right because you are charging for a product, and it has to be worth what the people pay for it.

For me, good bacon is a matter of pride: it's not just opening a bag of mix and applying it to the product and crossing your fingers. There is no substitute for experience and the more you cure, the more experience you get, the more you realise you don't know everything. Henrietta said to me at the end of the interview, 'I think you have a bit of magic in your fingers,' and we shook hands and she went on her way.

Some time later I had a call from a London television company which wanted me to take part in a television programme with a well-known chef. They wanted to come down and run a screen test to see if I was suitable. We agreed, and when the young researcher came with her team, she asked me to explain the business to her.

I explained about traditional bacon curing, using all natural ingredients with no additives or binders; smoking in a traditional kiln, all done without chemicals or artificial colours. The researcher said it was just what they were looking for. They shot some test film and said they would be in contact with me. In due course they rang to say they would like to come down and see me again.

The presenter of the programme was Sophie Grigson who came to introduce the first programme of the series from my farm. The reason they were interested in us was that everything we sold, we produced ourselves. In fact, we had a notice in the shop which said, 'We make all we sell and we sell all we make' and they thought that was unique.

When the day for filming arrived, we spent an hour on camera showing how the process worked with a focus on all the mixings – Sophie asked what each one was and what it was used for. They were also interested in the machines we were using, many of them unique to the bacon industry, such as a machine called an ET22 which sliced and stacked the bacon.

When we were first introduced, I thought Sophie was a very nervous lady, but as soon as the programme started, her professionalism took over and she was magnificent! Next she wanted to know about the smoke kiln. I started at the beginning and explained that we first dusted the bacon with a pea meal, to give it a natural, bright colour.

She also wanted to know which woods we used in the kiln and I explained all about the different flavours and colours from the different woods. Oak and beech, for example, both give quite a different smoke. We could affect the smoke by adding different herbs including coriander, mace and juniper. We had absolutely no mechanical parts in the smoke house, and smoked purely using our own skill.

After shooting in the smoke house, Sophie Grigson wanted to taste the bacon, so we took several samples up to the farm kitchen and she started to cook. She was amazed that the bacon did not shrink, and I explained that this was because there was no water in it; there never is in traditional bacon curing, whereas in the modern bacon curing industry, an additive called polyphosphate is added specifically to hold the water in the bacon thus making it heavier and more expensive by weight, so that you are paying for water, but we did not use polyphosphate as we did not agree with that.

All in all, they said it was a good shoot: the producer was happy, Sophie was happy and I enjoyed working with her. We had struck up a nice relationship; she was a nice down-to-earth lady.

The next stage of the programme was to feature a man called Graham Portwine, a gentleman in the meat trade. I aleady knew and admired him. The name Portwine originates from the French Huguenots, who had been in the Seven Dials area near St Giles Church, London, since about 1760. Graham ran Portwine's butcher's shop which I should imagine was one of the oldest butcher's shops in London. Graham had learnt his trade from an early age, and he had learnt it well. His business in Covent Garden provided the best meat and game to all the hotels, restaurants, and aristocracy in London.

His family business had been supplying good meat for well over three hundred years and had an excellent reputation. In your work, you meet certain people and they leave you with a warm feeling: Graham is such a man. When I first met him, I realised he knew who he was, and his shop was a time capsule: it was clean, but it was like walking into another world. All the fixtures were still as they had always been. He used to deal with me as one of his suppliers in my specialist area. Over the years we became great friends and I shall always be proud that Graham and I were in the same TV series and that cemented a friendship which has carried on to the present time.

In life, everyone has a dream and in a lot of cases the dream never comes true. My wife Trisha's dream was to have a little shop to sell high-class ladies' garments and I was going to help her to achieve this aim if possible. Attached to the house was an old cheese store, a room which was used in earlier days to store the cheese from the farm. It was quite convenient, as one of the doors from the house led into the old cheese store and there was another door leading onto the yard which would make a good public entrance. So we started to strip the place out, put some new door frames in and laid a nice carpet.

Trisha decided she would sell all English-made woollens, which I thought was a good idea. The manufacturers came down to see us, offered to help with the design of the shop, and all going to plan. It was quite expensive and an unknown venture.

Eventually the shop was finished, all the fittings were in and it looked very nice, although we had spent a little bit too much money and we still had to think about money for the stock.

I had an old stamp album which I had inherited, so I decided I would sell that to raise some cash. I knew one of my bacon customers was a stamp collector so I decided to ask his advice. When the man next came into the shop I said to him, 'Arnold, I have a stamp album and I would appreciate your advice if you have any idea of its value.'

He came up to the house and had a look at the album. He took out his eye glass and looked at each one. Then he explained to me that, apparently, the stamps were very early colonial and worth about a thousand pounds.

'But don't rely on that, that is just my opinion,' he added.

In due course I went to the local county town and found a shop which dealt in stamps and clocks. I asked the young lady if the owner would be interested in buying my stamps. She fetched the owner. As with Arnold, out came the eye glass and he inspected the stamps, looking at them all very carefully.

He asked me where they had come from and I told him I inherited them many years ago and I was not a stamp collector, but I wanted to realise some money on them.

He said, 'I will give you £750 for them.'

I shook my head.

'No, I want a thousand pounds for them,' I replied.

'Is there a reason for that figure?' he asked me.

'Yes,' I said, 'There is a good reason.'

He leant over the counter and shook my hand.

'A thousand it is then.'

He wrote me a cheque and I thought, 'Well, this will help launch Trisha in the shop.' I went back and told her she had money to buy the stock and gave her the cheque for a thousand pounds.

When she asked how I had got the money, I replied, 'Ask no questions and you will be told no lies.'

Trisha laughed and then she started to buy all her stock. I felt satisfied that I might have made somebody else's dream come true.

CHAPTER 8
Sunshine and Dark Days

One day the girls called me and said a lady had come in the bacon shop and she wanted to test all our products. I went down quickly and introduced myself. She told me she was a Standards Officer and that she intended to check our products: in fact she had already bought a number of the products and was in the process of cutting them up and putting them into jars.

I said to her, 'It would have been polite if you had introduced yourself to me before you purchased the product,' but she replied that it was the way they did things.

I told her I thought it was a very underhand way of doing things, but she seemed to think I was not entitled to my opinion. There and then she decided we needed a thorough inspection and so she went through everything: she weighed and measured all the products. She picked up one of the packs of bacon and told me I should have the packs labelled with the weight, price, time and date. I told her I did not think that was necessary as I produced the product on-site and when the customer chose their pack, at point of sale it was weighed, so we didn't need to do that in advance.

She agreed I had a point but insisted I re-label and I had not enough money to take them on, to prove my point. I was very angry at her instructions as I thought our way of living was being threatened by pieces of paper and Brussels. We lost on that occasion and I felt very disillusioned. I was a small producer trying to make a living and trying to produce a good traditional product. We could just about afford to give our customers the best, at a price they could afford, but not if the so-called Standards people were going to stick their costly oar in.

One day I was working in the factory when I heard a tremendous bang and I knew something was wrong. I left the factory, crossed the yard and looked out across the fields. There was a huge lorry in clouds of smoke, more-or-less on its side and I knew it was serious. The lorry was loaded with planks of wood and these had been strewn all over the field, it had demolished the fence and I knew by the state of the lorry it was a serious situation.

Meanwhile all the cows in the field had decided, as they are nosey by nature, that they would come and have a look with me. I walked over to the lorry which was at a precarious angle and the driver still in it.

I knew things were not good because there was diesel everywhere. I tried to open the cab door but couldn't so I ran back to the factory and asked the ladies to dial 999 for an ambulance and fire engine. Meanwhile I grabbed a crow bar and returned to the field, by this time accompanied by the ladies and the handyman. I told the ladies to stand back while the handyman and I crow-barred the door open.

The man was unconscious in the cab and I was very concerned about the amount of diesel that was pouring from the lorry. We carefully lifted the man, supporting his head and neck, and gently laid him on the grass away from the lorry. He started to come round, said he felt a bit sick and was worried he would get the sack. I told him not to worry as the main thing was he was alive: the lorry, the wood and everything else could be sorted out and these things were just part of life.

The ambulance service and the fire brigade arrived and they took over and cleared everything up: there were a lot of people milling about. The drivers of the breakdown vehicle which came to move the lorry wanted to talk to the injured man and started cross-questioning him about credit cards and how it was all going to be paid for. I went over and told them to stop harassing the man as he had just had an accident, but the breakdown man looked at me and told me it was procedure.

I said, 'You can see the man is in shock: he can't answer your questions! At least leave him alone for a while – use your common-sense.'

The lorry had mown down the fence and the cows were beginning to take an interest in the road. I asked one of the firemen if he could help me put the cows into the next field but he told me it wasn't his job; I would have to find someone else to help me.

I said to him, 'Common courtesy has left you many years ago. Can't you see the silliness of not helping me? If the cows get on the road, they could cause another accident.'

But he was a stupid man and still would not help. Luckily the ladies came out and helped put the cows in the field. I thought to myself I had had enough excitement for one day. The task of lifting the lorry and shifting all the timber took a couple of days. It was a really nasty experience: all the fences had to be repaired but fortunately the lorry company paid for all that. I thought to myself: you never know what's round the corner.

A few weeks later the lorry driver called in to thank me and the staff for coming to his assistance which I thought was payment enough, a very nice thing to do, as most people live for the day and courtesy has long gone.

One day a man arrived with a metal detector. He had done his homework on the surrounding area and he thought on our land there had been a battle. With our permission he wanted to do some research and go over the land with the metal detector. I said I did not mind and he told me if he found anything we would split the find, half shares each.

Within three or four days he arrived with his metal detector and went all round the fields and he came back into the shop at dinnertime. He had found a lot of lead shot and he said there had been a lot of activity, mainly in the big field in the middle. He said it would mean digging some of the field, possibly very deep, and I asked him how deep did he mean? He said he did not really know but he thought about five feet down. I told him I thought that was a long way down to dig, and he agreed with me but he was very excited as he thought he had found something.

I agreed for him to dig an observation pit which he duly did. He had dug down about three feet when he found a cannonball, the cannonball I still have today. We took it back to the farm and washed the mud off and there it was, a perfect cannonball, still there after hundreds of years.

I too became very excited. I had visions of buried treasure and becoming rich and famous! He came back and said he thought whatever it was, there was a lot of it, so I suggested, as we had formed a partnership, the fairest way to do it was to jointly hire a digger and see what was there and split the find between us.

He thought that was fair but I said, 'Just before we order the digger, let's be sure: go and get another sounding.'

He put on his earphones and there was a tremendous noise and I said, 'There is definitely something there. Whether it's old pipes or metal, we will never know unless we dig.'

He agreed with me, so no more to do, we engaged a digger.

The digger arrived. It was a lovely day and we were all excited about the treasure we were about to unearth. You never know in life what is round the corner. The digger-man started to delve into the soil and the first clods came up. The driver suggested he dig in rows, at a depth of four or five feet. He went down and there was something there! He stopped the digger, the metal detector man jumped into the hole and pulled out the object and put it on the side of the trench. It was a harness off a horse, the old fashioned harness with two spikes sticking out either side. We took it back to the farm and power-hosed it off. It wasn't old, probably Victorian, and it was in good condition considering it had been in the ground all those years.

We carried on digging and we unearthed about thirty saddles and bridles. We went through them; most of them had perished but the brass was in quite good condition and the metal detector man decided to keep the brass so we divided this up between us.

The metal detector man took his to auction and was well rewarded for his day's work. To this day I still have the cannonball and some of the brass medals from the harnesses; we had no riches from that exciting day but it was a very enjoyable dig which lives long in the memory.

You never know what the future will bring, so each new day you live, you should treat as a bonus. The secret of life is to enjoy as much happiness as you can because each day is precious, and no matter how wealthy you are, you can never buy that day back. That is the lesson I have learnt over the years: to enjoy life to the full.

On one occasion we had had a long and hard day and we had gone to bed exhausted. During the night Trisha awoke and on her way to the toilet, tripped and fell over a chair. This was about three o'clock in the morning and I knew it was a serious matter. I called my daughters Anna and Rachel and we realised Trisha had a serious injury.

We made her comfortable, called the Doctor but she would not come out and told us to call an ambulance instead. I was very cross because when you need help, you need it, and in this instance it was not forthcoming. Still, the ambulance arrived very quickly, the Paramedics examined Trisha and said she would need to be taken to hospital, as she was in terrible pain. They managed to get her into the ambulance; Rachel went with Trisha and I followed behind with Anna.

Trisha was taken into Accident and Emergency where she was examined, an x-ray taken and it was found she had fractured her femur. I asked for her to be put in a private bed and for everything possible to be done. We went to the ward and saw Trisha settled in. By this time she had had some pain killers and was feeling slightly more comfortable, so we left her to sleep.

Anna, Rachel and I made our way home in the dawn, all in shock and worried by the night's events. The hospital rang to say they where going to operate on Trisha straightaway to pin her femur and we could visit the following evening as she would be in theatre for a while and would be feeling very groggy when she woke up. When the evening arrived, Anna and I made our way to the hospital and found Trisha very chirpy and feeling a lot better. The week progressed and our daughters visited and generally spoilt Trisha.

The telephone rang one afternoon and the ward Sister asked me to come in quickly as Trisha was not very well; she had a blood clot and was going to be put on a drug to get rid of it. She said the situation was serious but I did not realise how serious.

As I was driving along towards the hospital the traffic held me up and I glanced at my watch and it was ten minutes to five. I knew at that moment something terrible had happened. I arrived at the hospital, parked the car and rushed to the ward but the Sister came to me, took me into her office and told me very gently that Trisha had collapsed and died very suddenly. I asked what time she had died and she said it was at ten minutes to five.

At that moment I knew I had lost my friend, my lover and my mate. I felt very, very lonely and life would never be the same again. For the first time in my life, I had no answers, no plans, just a huge void and I felt very desolate and alone.

The next week passed in a blur. All the procedures when someone dies had to be gone through and to be frank I did not know whether I was taking part or not, as it all seemed so unreal. After the funeral I went back to work, but life was not the same. I woke up each morning and went through the motions of the day; it was as if my body was there but not the rest of me.

One day the telephone rang and it was the Council, a man from the records office wanted to speak to me and wanted all the details. I said I did not really feel like giving any details at the moment, I wanted to leave things for a while.

He said, 'What's the matter with you? Everybody has bereavements in the family: buck up and face up to it.'

I was so angry I said to him, 'Sir, I hope you never feel as bad as I feel now,' and put the phone down.

I thought how inconsiderate people can be at moments of great sorrow.

CHAPTER 9

Gypsy Bacon and a Bargain Riley

Time passed. I felt I was living in a bubble and not really taking part. Work filled my day but the evenings seemed relentless. I drank too much and smoked far too much but nothing took the pain away of losing the person I loved.

I was called to the telephone one day: a lady called Winifred was asking for me. She said, 'We have a smallholding in Hereford and we cure fat bacon for the Romany community and we smoke the bacon in the very old-fashioned way in a barrel but unfortunately we have had one or two barrels catching fire. We were wondering if you would be interested in building a traditional smoke house as I heard you build traditional kilns.'

I told her I would come and have a look. I told her I was one of the last traditional smokehouse builders in England, and that I built them traditionally, as they would have been built two hundred years ago.

She gave me complicated directions on how to find the farm; she told me, once I had turned off the main road, she would put four yellow buckets along the country road and I was to follow these until I reached the fourth one and then to turn right up the lane. In due course I came off the main road and, true to her word, there were the yellow buckets.

I followed them until I came to the fourth one and then turned up the track for the farm, a narrow track, but I took my time and I

106

arrived. It was a typical Herefordshire farmhouse, black and white, which had been extended over time. Winifred came to the door, greeted me and told me her husband's name was Alfred.

After all the introductions, she explained, 'We have been producing this bacon for the true Romany gypsies, for many years. My father used to produce the bacon for the travelling community so it goes back a long time, but we have a problem at the moment. The gypsies do like some of the bacon to be smoked and we are finding smoking the old-fashioned way, using the barrel with sticks across it, is causing one or two fires and we want to prevent this.'

I replied, 'Yes, that is one of the problems with using a barrel. You can get away with using a barrel for small quantities of bacon, but if you use it every day, the fat begins to build up round the barrel and if it gets too hot, you then have spontaneous combustion and frequent fires. If you use the barrel system on a regular basis, you can't get away with it'.

Winifred asked me to come and see how they had converted the dairy into a curing house so I followed her into it and it was scrupulously clean. Hanging up were huge pieces of fat bacon and I had never seen such fat bacon before.

She said to me, 'This is what we do: we raise the pigs and make fat bacon specially for the gypsies and travelling people. It is a very profitable business, and that is all we do on the curing front.'

I told her I thought they were unique.

Alfred her husband told me another line to their business was the renovation of the old gypsy carts. I asked him to show me and we went into the barn where he was restoring an old gypsy caravan and it was absolutely beautiful, a complete work of art.

I was so intrigued by the way he was painting it, all the flowers and roses, and I asked him how he did that. He showed me the moulds and how he painted the roses and the different designs which were all hung up with different numbers on, for him to copy. He showed me how he repaired the wheels in the old way, on the ground, putting the rims on, heating the rim to make it fit. It was a really interesting day out and I felt privileged to have a peep into his unique business before it disappeared forever.

Winifred asked me if I would like to look at the pigs, and I said I would love to, so off we went to one of the fields. The fields were very well fenced and as we approached, Winifred called out and it was as if the pigs had been offered a day at the seaside: they all rushed up to the fence.

I was so taken aback! The pigs were all the size of small ponies; I had never in my life seen pigs as big as they were. Winifred told me they were four to five years old and that is how they achieved the very fat bacon. Usually pigs are fully grown by about seven or eight months old and I had never seen pigs that large before. I remarked at the way they all came when Winifred called them.

'Oh yes, they all come, but we never name them because then it would become personal and it would be too hard to take them to the slaughter house.'

I thought to myself, 'Well, Maynard, this is a lovely day out and something you will always have to remember.'

As we stood there, a very large, very smart vehicle arrived and a gentleman alighted. You could tell from his stature, the gold sovereign ring, the huge gold necklace and the red neckerchief, that he was from the travelling community. They all greeted each other very affectionately and he told them he had come for some of the bacon. We all went into the shippon and Winifred pointed out the bacon. There was some twelve months old, some six months old.

He went through the bacon like he was choosing a fine old wine, then Winifred put muslin cloth on the pieces, weighed them out and gave him the bill – it was £600 and I thought to myself, 'Well, they make a better deal than I do but good fortune to them!' He paid the bill and said he would see them in a couple of months.

I asked him about the bacon he had just bought, and he told me it was the only bacon that would keep in the caravans, as it was as hard as a brick and dry. He told me it was a good investment and I thought to myself, 'Horses for courses.'

That was a part of the industry I had never seen before and you never stop learning; true wisdom comes when you realise you still have a lot to learn. I went back into the shippon and Winifred said, 'Well that is another satisfied customer.'

I asked her if she only dealt with the travelling people and she said, 'I do and I don't but it is predominately travelling people, as we cure the hard bacon especially for them.'

I said, 'You only use salt?' and she said that was the case and she showed me her salt. I looked at it and said to her it looked a good salt, but had she tried a larger grain salt? That often improved flavour and she thought she would try that.

It was time to give Winifred the advice she had asked me for about the smokehouse. The first thing to do on the farm was to find where to site the smokehouse. I explained that siting a smoke house was like sailing a ship: it was all dependent on the wind. I told her I would be very grateful if I could walk around for about half an hour on my own and I would then tell them the best place to site the smoke house.

She agreed and told me to go up to the house when I had finished. I stood for a bit and looked around and then walked about. I was looking for the northerly wind. I took my handkerchief out of my pocket and held it in the breeze. It is important where the smoke oven is placed, because it has no moving parts and it needs to face the prevailing draught. You can control the amount of draught with the fins in the chimney. It is very simple; like most simple things it has stood the test of time. I carried out one or two experiments, I marked it out and went up to the house, knocked on the back door was invited in.

I went into the kitchen, a very long room with a great big grate at one end with two huge oven doors and a fire in the middle. It must have been one of the earliest Coalport grates. Winifred had looked after it and kept it well polished and I knew these people were happy in their own environment. There was a nice table with some fine Windsor chairs and some lovely corner cupboards, but not on show, they were part of their living tools and everything in the kitchen was used. It was a working farmhouse kitchen and everything in the kitchen had a purpose.

Alfred asked me where I thought they should build the smoke house; I told them I had marked it out roughly in the yard and if I had some paper I would draw the smoke house design for them. Winifred found me a large piece of paper and Alfred cleared a space on the table by simply sweeping things onto the floor!

I kept my face straight and sat down and sketched the smoke house for them. I explained how to put the bed, or base, down for it, and the safety feature on it, in case a fire broke out in the smoke house. They were intelligent people and followed my every thought. The Grandfather clock in the corner started to strike.

It was about four o'clock; it struck the four chimes – but then carried on chiming, so Alfred stood up, went over to the clock and kicked it! He said, 'I will have to get that fixed!' I had a secret smile to myself, as I thought that was really funny.

I was sitting in the chair and felt a funny sliding sensation go over my feet, I looked down and believe it or not, there was a snake! I must have jumped five foot in the air. I have never been so frightened in all my life, as I do not like snakes and have always been frightened of them.

They both laughed, 'Don't take any notice, that is only Harold, our pet grass snake!'

By this time I was sitting back in the chair but with my feet held firmly off the floor. I'd never been in a house with a pet snake before and I thought the best thing to do was to stand up and if I saw it again, it would be easier to get out the door. I wondered what else I would encounter.

I went through all the stages of the building of the smoke house, where to put the vents and how to do the first smoke. I told them how to deal with any more fires. It is an old curer's trick: you need a bucket of fine sand and a bucket of fine salt and you mix them evenly, half and half, so you have two buckets of the mix. This you throw on the fire and it immediately douses it. The old curers and the smokers have used that method for centuries.

I had had a wonderful day – despite the snake – and when it came time to leave, Winifred asked me what they owed me and I said, 'I think £250 would be reasonable.'

Winifred said, 'I think that is very reasonable, Maynard, you have given us a lot of information.'

She went to the dresser and took out a large box and opened it and I had never seen so much money! She paid me there and then. I thanked them for their kindness and I knew I had met some rare and very nice country people.

Walking across the yard one day I noticed a vehicle coming down the drive. The driver parked the vehicle and three men and a lady alighted from the car. One of the men, short and with glasses, asked to speak to Maynard Davies.

I told him that was me and he replied, 'We've come to do a spot-check on your diesel fuel and on the books. We are a new department combined with the Inland Revenue and the Customs and Excise.'

I asked him, 'Why have you given me the pleasure of your company?'

He replied, 'It's a random search. We are a new team of investigators and this is what we do.'

'You do realise I am trying to run a business here. Unlike your organisation, which is paid for by the tax-payer, I do not have any spare labour to deal with you. You should have made an appointment.'

He looked at me. 'We don't do that. We have full authority to go anywhere in the country without a search warrant.'

I muttered, 'Even Buckingham Palace?'

'Yes,' he replied, 'Even there.'

I asked them exactly what they wanted to look at and he informed me they wanted to look at all the paper work and dip all the diesel car tanks because they were looking for the illegal use of red diesel (we did have a red diesel tank for the generator and as you know, red diesel should only be used on farm equipment, agriculture or generators). They told me they wanted to dip the vehicles and they wanted to dip mine first.

The boss told me he had the number of my truck and I thought to myself they had done their homework well. We went round to where I kept my truck, he put what looked like a large syringe in the tank and siphoned some diesel up.

He said, 'That looks clear, but I have another test to do.'

I was beginning to feel very annoyed and frustrated. 'I want you to hurry up because I have got a lot of work to do and I have not got time to mess with people like you, wasting my time.'

This very officious short man pulled himself up to maximum height. 'You do realise who we are?'

Exasperated beyond measure I said to the two who had accompanied me, 'Yes I know who you are, you are a couple of pratts!'

I knew I was losing my temper. He started to read off this piece of paper but a red mist came in front of my eyes. I knew we were straight people and I had done nothing wrong.

I hit the side of his vehicle and dented it.

'I advise you very seriously to get the test done and then sit in that car.'

I think then they realised they had met somebody who was not going to be walked over. They finished the test and then sat in the car. Meanwhile the lady and the other gentleman waited for me to go into the office. My office was only small with a desk and a filing cabinet and it contained no mystery, it was straightforward. We were not making a fortune, just a living. I thought to myself:

these four people here think they are important and then a bit more! They asked to check all the records. I did point out to them I did not have full time office staff, only a part-time lady and she was not in that day.

My old dog was a grumpy old dog; I knew his temperament was variable as over the years he had done some naughty things and it had generally cost me a full ham or several packets of bacon to soften the blow of having been nipped by Brucie. He had got a reputation of nipping before barking.

I knew his favourite place was to lie in the office by my chair when I was working. I thought to myself, 'Opportunity knocks very softly.' I stood back and let them enter the office ahead of me and shut the door: there was a scream of fear and a shout for help – I counted up to twenty and then opened the door. The lady was standing on the desk and the very officious gentleman was holding a waste paper basket in front of him for protection. Brucie had got them trapped.

I asked in all innocence, 'What's the matter?'

The man spluttered, 'Do you realise you have got a very dangerous animal there? Do you realise he attacked us?'

I tried to look concerned.

'He was only defending his home. When you came with that very important piece of paper telling me you had access to everywhere, that was correct, but Brucie he did not understand that, he did not know how important your piece of paper was,' I said.

The man puffed his chest out.

'He has ripped my trousers.'

Still trying to look contrite I said, 'Well, he normally does more damage than that; you've got off very lightly.'

Surprisingly they did not stop much longer and he told me he would put on his report that I had a dangerous animal and I was very uncooperative, that the visit was not completed and they had not inspected all they had come to see.

I told them they would be very welcome to come back at another time but I knew that would be last I would see of them. They had come and found nothing and done nothing.

In the food industry there are certain jobs you dread, the same as in any job. One of worst jobs was to clean out the smoke house flue: the smoke travels along the flues and they become blocked and need cleaning out. A whole day was best put aside for this job, a day where you had plenty of time and plenty of patience, so one Monday morning I decided to do this.

I went into the smoke house and started to unscrew all the vents and get the brushes into the flues. The residue came out as very fine silk, and that was the remainder of all the smoke. Occasionally you did need to go into the smoke oven with a hard brush and brush the walls down, because scale could develop on the walls.

This was all part of general maintenance, but you never really cleaned the smoke house as such, because the flavour was in the bricks. But the maintenance used to take the best part of a day and when I had finished I would be totally black so I used to wear an old boiler suit and an old hat to do this job. Afterwards you needed a thorough scrubbing to take all the black off, and the clothes usually had to be thrown away.

On one occasion one of the flue brushes broke and I cussed a little bit, and decided to go and fetch another one. I got into the truck and went to the nearest town to buy one. Driving along the road I happened to pass a car lot, and there on the forecourt was a vehicle I was looking for, a lovely Riley, a lovely maroon Riley. I drew into the car lot and in the office was a salesmen sitting in a leather chair.

I said, 'The vehicle over there with the price on: is that the correct price?'

'Yes,' he replied, 'but you won't be able to afford it.'

He had obviously looked at my vehicle, which had seen better days and at me looking like a pauper, but that was his poor judgement! I went over to the Riley and had a good look at her. I went back to the salesmen and asked if I could have a spin in her. He replied, 'Not really, you can't have a spin down the road in that state; you will knock hundreds of pounds straight off the vehicle.'

'Right, Sir,' I said, 'What will you give me for my vehicle in part exchange?' He wrote some figures down and pointed at them and said, 'I will give you that, for your vehicle.'

'You'll give me that figure for my vehicle in part exchange?'

'Yes, I will,' he replied.

'Okay I'll take it!' He nearly fell out of the chair, he was a huge man who must have weighed twenty stone. He repeated, 'You'll take it?'

'Yes, I will,' I replied. 'Furthermore, I have sufficient cash with me to do the deal right now.' I thought he was going to have a heart attack – the look on his face was a picture.

He said, 'Now?'

'Oh yes,' I replied. 'Now.'

I had sufficient money to do the deal. I knew it was the moment because he had quoted me twice the price my old vehicle was worth. I paid him there and then in cash. I sensed a reluctance on his part to give me the keys, but a deal was a deal. I gave him my name and address, fetched all my bits and pieces out of my old vehicle and left it there.

I carried on with my journey, went into the town and bought myself a flue brush, and I thought to myself, 'Well, it was not a bad deal and the flue-cleaning day hasn't turned out too bad after all.'

When I arrived home about an hour later, the phone rang and it was the sales manager asking me if I was the gentlemen who had bought the car off the forecourt and had left the other vehicle behind. I confirmed this was so.

He said, 'Well, my salesman has made a faux pas. The vehicle you left, he has given you twice as much for it as it is worth.'

'Well on this occasion,' I replied, 'You will have to live with your mistake.'

The sales manager tried a new tack.

'Well, we won't change the vehicle over with the licensing people,' he said.

'That is up to you, and if your salesman had managed to hoist himself out of his chair instead of swivelling round to me and making a snap judgement on me, and if he had looked at my vehicle properly, you would not be in this situation. I'm afraid I am going to make a stand, and it won't harm you lot to learn a lesson. Its not often in life that anyone wins with the second-hand motor trade! So on this occasion, I am going to stick by the arrangements.'

I knew by his tone he was defeated. He really had made a mistake and the old saying, 'You should not judge a book by its cover' rings true on this occasion.

CHAPTER 10

A Light Shines

I was busy working in the shop one day when a lady came up to the counter and she said that she wanted some very nice bacon. I told her we had about twenty different varieties that day and she asked me about them. I showed her one of the milder ones and she decided she would have that one and then she looked at the eggs and asked me if they were my own chickens and I told her they were. She told me she kept chickens and we had a discussion on the merits of chicken keeping.

I asked her if she sold them and she said she took them to work for her colleagues as she was a nurse working in the local hospital. She asked me how many eggs a day I got from each chicken and I told her one a day, but she said she thought some of her chickens laid two!

I commented, 'I think one is a good day's work for any hen,' and she laughed. She told me she would go off and try the bacon and promised she would come back and tell me if she had enjoyed it.

A few weeks passed and she came back into the shop. She said she had enjoyed the bacon and that she would try another cure. We chatted for a while as the shop was quiet, and we seemed to get on very well. I suppose in life there are some people that you just click with: whether its personality, attraction, genes or whatever, there are other kindred spirits you feel comfortable with and then there

117

is the other side of the coin, you meet people you get an uneasy feeling with who are bad people and I suppose its all down to your senses. I had a lovely feeling with this lady and we talked a lot and she told me she was on her own and she asked me if I was on my own and I told her I was.

I said to her, 'Anyway, I could take you out for a cheap meal if you would like.'

She smiled at me.

'A cheap meal?'

'Oh yes, a cheap meal.'

'I'll take you up on that.'

So we arranged to meet in the restaurant's car park. When she arrived, the most striking thing I noticed was that she had the most wonderful hair. When she had come into the shop it had been tied back into a bun, but she had very long fair hair and it had been left to flow round her face. I asked her if she was a natural blonde and she said she was.

We made our way into the restaurant and sat down. I ordered us some drinks and while we perused the menu I asked her about her life. She told me she was a Nursing Sister in the Coronary Care unit where she had worked for a number of years. She lived on her own in a cottage in the hills and she had for company two dogs and the chickens. I told her my situation and that way we both started off with honesty. We had ordered our meals and eventually they came; she really enjoyed her food and had a good appetite.

I jokingly said, 'I'd rather keep you for a week than a fortnight,' and she laughed, we both felt comfortable with each other.

We arranged to meet again and that was the start of something special which still continues today. Our relationship progressed, my youngest daughter decided to leave home to start her nurse training so Ann and I decided to set up home together.

Sometimes milestone events stay forever in your memory: I remember the day Ann came to set up home with me. She arrived with two cats and the most wonderful dog I had ever seen in my life, named Dilly. I had had a lot of animals in my time and had sometimes found they were as good company as people, and more reliable, but this dog Dilly was outstanding. Dilly and I struck up a rapport straightaway. I had a dog called Lassie, a small Alsatian, and as life has some funny quirks, Dilly took to me and Lassie decided to adopt Ann and that was the situation; I don't think there was any hard feeling on either side.

Lassie used to follow Ann around all of the time and guard her, even by sitting outside the toilet! Dilly in turn followed me everywhere, sitting outside the factory for hours on end, so all of us cemented our relationships. Dilly had very acute hearing and one evening we were all sitting round when she started to howl. I thought there wasn't anybody about. I looked out of the window and could see nothing. I told her she was a good girl but Dilly scratched at the door to go out but I told her it was only the rabbits in the field: but I was wrong.

The next morning we went down to the factory and immediately I saw the bolts on the door had been sheared off. I walked into the shop and saw a lot of the hams had gone. I went into the factory and all the doors of the fridges were open and a lot of bacon had gone too.

It was near Christmas and the burglars had gone through everything and turned it all out – it was a terrible mess. The float in the till had gone: I used to leave forty or fifty pounds in change to get us going at the start of the day. They knew what they were about. I rang the police and within half an hour they had arrived, and quickly established that the thieves had come across the fields at the back of the shop.

I said, 'I don't think that is possible.'

'Yes,' said the policeman, 'It is. Your front gates are still secured by a large lock and chain, so entry is limited – they have driven across the fields.'

The old Sergeant told me to follow him and we walked down to the field and saw where they had taken a section of the fence out, then brought luminous balls and laid them all across the field, every fifteen feet, so they did not need to put their car lights on, just followed the balls to guide them. They had taken a good proportion of the stock, so it was a very difficult time for me.

I asked the Sergeant what would be the chances of catching them and he said, 'Not a chance. They knew what they were doing. It will all be on market stalls by now.'

He asked me how much stock they had taken; I told him I would have to do a complete stock check to give him an accurate figure,

The next thing to do was to make the place secure as all the locks had been shorn off the doors. I rang the locksmith and he came out and told me he would put a huge lock and chain on the front gate which nobody would ever get off. I told him that was unnecessary. He said such a lock would cost me a hundred and fifty pounds and I told him that was far too expensive.

He said no burglar could ever break it but I said, 'It might not be broken by the burglars but it would break my wallet!'

He thought he was going to make hay whilst the sun shone. He charged a tremendous amount of money for putting the locks on and up to then I did not think the day was turning out very well. I'd gone through this experience before but at least I thought we were well insured and believed at least I would recoup most of the losses. I rang the insurance company and they said they would send an assessor down and asked if I had any idea of the amount of stock lost. I told them by the time the assessor arrived, I would have a complete stock check for them.

A date was arranged and the insurance lady arrived. I offered her a cup of coffee and she accepted. I showed her the scene of the crime; she inspected every door and window, paying particular attention to all the locks and I knew by her demeanour that the company was looking for a loop-hole not to pay out. She said all the windows were well-secured and the doors had good locks.

She asked me how much we had lost and when I gave her the figure from the stock check, she looked at it and said, 'You are only insured for eight thousand pounds but you are saying fifteen thousand pounds' worth was taken?'

I told her that was correct, but pointed out to her that at this time of the year, we carry far more stock as it was Christmas time, but in the summer we would only carry a thousand pounds' worth of stock. She said to me my claim was not valid because I was under-insured and that she couldn't pay out.

I said to her, 'I don't look at it like that, I think that is fraud. You have taken my money over the years and now I have had a burglary, you don't want to pay out. If I had had a burglary in January when I had very little stock, would you pay out then?'

She replied, 'I can only deal with the present circumstances.'

So in the end, the insurance was void. We had to bear the cost of the complete loss, the locks, the fences, the loss of stock which ran into a considerable amount of money. The insurance was not worth the paper it was written on. From that day on I have always viewed any insurance with great suspicion. We ended up well and truly out of pocket because on a technicality they would not pay out. So heed this story and always read the small print. All in all, the day was not one of my better days.

A man rang me from London and told me he was a supplier for some of the major hotels, shops and restaurants and would I be interested in supplying him, as he had tasted our bacon and he thought it was a good product? He told me it would be a specialist order and asked if we had the skills to fulfil his requirements.

I told him the best thing for him to do was to come and pay us a visit and see for himself. He came down and he said that he was a specialist supplier to the top hotels and he said some of the orders were very unusual. Could I, for instance, do a London smoke?

I told him I could as I done had this for many years when I was an apprentice in the Black Country for Thea. He had had an order for a large establishment in the city. We had done a London smoke many years ago and I remembered exactly how we did it. It was done basically with beech wood, barley straw, all in equal amounts, and we also added coriander and caraway seeds to the smoke. It was a lovely smoke: the beech and the barley straw gave the bacon a rich golden colour. The London smoke was smoked rather longer than I normally smoke bacon, because it does not go too dark. The gentleman was surprised I knew all about the London smoke: he hadn't expected that.

Next he asked if I could do the Ayrshire roll and I asked him which Ayrshire roll did he want me to do: the one with herbs or the one without? He told me he did not know so I told him he would need to make up his mind about that. Next he asked if I could do the original Canadian bacon? I told him I could cure the original Canadian bacon which is done with pure Maple syrup, but I would need definite specification on exactly which products he wanted.

He seemed to be impressed with our product list. I told him we could produce his order for him. Looking at the man, I could not think who he reminded me of, and then it came to me! He reminded me of Guy Fawkes with his pointed beard! He gave me the order with all his specifications: it was a huge order and he asked me how long it would take? I told him it would take a couple of months before everything was ready, I would ring him when the time came and he could come and fetch the order.

We carried on with production and I produced the London smoke. I obtained some peat and barley straw and made a lovely smoke and I really enjoyed doing this one. When it was smoking with the coriander seeds and the caraway, as you opened the door, you had a lovely aroma.

The oven was by now maturing beautifully, as when you first build a smoke oven it takes a long time for the aroma to build up. The aroma is in the tar on the walls which is why you never really scrub a smoke oven out.

We put all his order together and he arrived and asked to test some of the products. I told him that was fine. He selected one of the products, I sliced some pieces off and cooked them and he was extremely pleased. He said it was excellent. The bacon did look very nice and I was pleased with how the order had turned out, because in our business, it is a challenge to do the thing right and I was proud we could produce things that were difficult to produce. This order was not run-of-the-mill stuff: you needed a little bit of nouse and a lot of skill to produce this specialist bacon.

We loaded it all into his truck and the time came for the tally to be paid. He took out his large cheque book and paid the bill which came to about five thousand pounds. I banked the cheque and in due course the bank rang up and said there were no funds available for the cheque to be cleared and I knew then we were in a very difficult position. I had the buyer's telephone number. In the bacon industry everyone starts very early in the morning, so at seven o'clock I rang him. I was lucky because he picked up the phone up.

I said, 'Maynard here. We've banked your cheque but it's bounced.'

'Oh, that's a minor matter; I have just had a little cash flow problem. I've plenty of cash coming into the system, so if you re-present the cheque again, it will go through.'

I said, 'I sincerely hope so.'

He replied, 'If you have any fears about my solvency I could come down for another order and I will pay you cash.'

I thought to myself: that will suit me fine. He told me to do half the order again and he would come and pick it up and pay for it in cash. I told him I would not re-present his earlier cheque: he could pay me the full amount for both orders when he came. He agreed and said he had had some very feedback about the bacon. I put the phone down feeling a lot better and thought to myself: it was just a misunderstanding; it was one of those things.

True to his word, he arrived to pick up his order with his pointed beard and his dicky bow. Again we loaded the truck and he complimented us on a good product.

Then he said, 'Right I will settle this account in full,' and I thought to myself, 'Manna from heaven!'

He said, 'By the way, Maynard, before I forget: have you any publicity material? Some of the customers are aching to put up notices saying where the bacon has come from.'

'Yes,' I replied, 'I'll just nip up to the office and fetch some.'

He nodded his head, 'Good chap.'

It took me a while to find some posters and advertising material and when I came back to the shop the staff told me he had gone. He'd wished them Cheerio and driven off. I had to go back to the house and make myself a cup of tea. I knew I had really been had and the chances of getting the money were very bleak. I knew one or two people in bacon in London, so I rang them up and asked them if they knew this man.

'Oh yes,' they told me.

I asked if they thought there was any way of my bill being settled and was informed he would make himself bankrupt and there would be no chance of me having my bill settled. Not to be deterred I rang his number. The phone was answered by a man and I asked to speak to my customer but was told he was not there. He explained I was speaking to one of the official receivers.

I explained I was owed quite a bit of money by the company they had now taken into receivership, and he told me I could put a claim in but it would not be worth the paper it was written on: everything was on lease, even the pens, so there were no assets. He said this man would get away with it, it was all a scam, and he would be trading again tomorrow with other premises and the name on the van changed.

That was a very hard lesson to get over and since, I have never liked men with pointed beards and dicky bows! I suppose there are gentlemen amongst gentlemen; and then there are gentlemen – and everyone has to decide which one they want to be.

One day in the factory the girls said to me the Hygiene Officer from the Council was in my office and would like to have a word with me. I knew this was going to be a pain, but all credit to myself, I kept calm. I walked into the office and greeted the Hygiene Officer and asked how I could help. He told me he had come to inspect the premises, a full inspection, to see how things operated and he told me he had never inspected a business like this before.

He said to me, 'I believe you have a smoke oven,' and I told him I had. I knew this was going to be painful as he had clutched under his arm a brand new brief case and was wearing a very shiny pair of new shoes. This wasn't going to be fun.

He asked me if there was anywhere to change and I showed him the room where the ladies changed themselves and said I was sure they would not mind if he used a little bit of space in their room. He looked at me and did not know whether to laugh or make a comment. I knew this was going to be a long day because people who have no sense of humour are hard work.

This man was in a new job and he was unsure of himself which made the situation worse. I had been visited by inspectors all my life and I just knew this one was going to be difficult.

We started off in the factory. Straightaway he said, 'The fly catchers you have here, one up this end and one at the other end, are the wrong way round: they should be facing the door.'

I told him I would rectify that but privately thought it was crazy, as I am sure the flies weren't too bothered if they were facing the door or not when they were zapped!

When a fly catcher is turned on, a blue light comes on and the flies are attracted to the light, go in and are electrocuted. Which way they are facing is irrelevant!

The next thing he wanted to do was to inspect the fridges. We had a number of fridges for different purposes. I explained what each fridge was used for. We went into the fresh meat fridge and I explained exactly what temperature it was set at. He said he was concerned the floors were uneven, and I explained they had been

125

designed to slope so that all the water runs to the drains and does not lie in puddles. Basically all factories in the bacon industry are designed this way as we use a lot of water in the curing process. He accepted that.

Next we went into the biggest walk-in fridge and he spotted that there were no lights in the fridge. I told him that in traditional curing, the bacon is not kept in the light because light affects the curing process. We only put the light on when we were working in there for long periods.

'Oh no,' he said, 'I want more lights than this in future, so I can inspect everything.'

I agreed with him then and later did nothing!

Then he carried on inspecting all the machines. On one of the machines he said there was only a small turn-off button but I pointed out to him the 'Dead Button' on the wall to turn off the electricity to all the machines for cleaning purposes.

He went up to the slicing machine and said, 'This is a very dangerous machine.'

I told him they had been in operation since 1909 and I didn't think anyone had ever had a fatal accident using one. I don't think he knew whether to laugh or cry: he could not find anything else to look at; he had given the factory a thorough inspection and could find no fault. He asked me if we used protective gloves and aprons and I told him we did. I pointed out the hand-washing sinks and the preparation sinks.

He asked me how long I had been in the business and I told him we had been at it for fifty years and that in all that time we had always had a clean bill of health. In the curing business, you can't be dirty, or you infect the product, so I told him the first rule was always cleanliness. Then he had a brainwave.

'Where is your first aid kit?' he asked.

I pointed to the big red cross on the wall behind him and one on the other side of the factory and told him we actually had two first aid kits! He then wanted to have a look inside them. They had never been used and were in pristine condition, so once again he was foiled!

Leading into the factory were large double doors. He asked if they were always kept shut and I told him they were kept shut most of the time but there was a very efficient metal fly curtain for the occasions they were open.

He then wanted to inspect the drains outside, so outside we went. My old dog Dilly always sat outside the factory all day long; she was my special dog, with a mind of her own and very friendly. She greeted all the customers with much tail wagging, and had been known to sneak into the factory and steal a string of sausages when the urge took her!

She always came down to the factory first thing in the morning and sat outside all day long, come hail or sunshine, as part of her ritual. So when I took the Inspector outside, Dilly came up to him to say 'Hello,' but I could tell he was not an animal lover and Dilly sensed this too and went and sat down in her allotted space.

He wanted us to lift the drain covers up. I did this for him and he said with glee, 'You have not got a grease trap!'

'Oh yes we have,' I replied and I pointed it out to him.

A grease trap is a container in the drain which collects all the bits and all the fat and it is cleaned out about once a week. It saves the fat going into the sewage. He seemed very disappointed that he could find nothing wrong, but he had a last card to play.

He wanted to look at the smoke oven, so off we went across the yard to the smoke oven.

He looked at it and announced, 'If anyone got locked in the smoke house, it would be fatal.'

I did not think this was likely to happen, but I told him I supposed anything could happen – we could even win the lottery – he was not amused. He asked what would happen if a fire broke out in the smoke house and I told him that that eventuality had been dealt with, in the shape of two red buckets, filled with sand and salt, which would extinguish any smoke house fire. If there was a fire, we would use the buckets then shut the vents down and that would automatically stop the smoke.

He pondered all this information for a minute and then said: 'I want you to cut a hole in the door so if anyone is accidentally shut in, they could release themselves by putting their hand through the gap to release the handle and they would be safe.'

I agreed with him to do that and decided to do nothing! I asked him if he had seen everything he needed and he said he had. He said he would not write to me as he was satisfied, so I thanked him.

He then went back to his car. Dilly tried to make a fuss of him but he was still not interested, so to show no hard feelings about ignoring her, she peed on his car!

I went back into the factory and got on with work but about half an hour later, one of the girls came to tell me the inspector was still sitting in his car, watching Dilly, who was as usual sitting outside the shop waiting for me. I looked outside and sure enough he was still there. I thought he was probably filling forms in and ignored it for a while longer, but when I looked out again, he was still there,. I thought he must have very long forms!

Dilly meanwhile was still in her usual place, doing her job as she thought fit. Another half hour passed and I looked out again – the inspector was still there. It suddenly dawned on me that he was waiting for Dilly to go in the factory and then he would have achieved his goal – he would at last have caught us out!

An hour had passed. The girls made a cup of tea and I told them to take one out to the man in the car, which they duly did. He declined it and drove off promptly. Dilly watched the car go down the drive and turn onto the road. She stood up, shook herself and strolled into the factory to see me, as if to say: 'He's gone, Maynard!'

My daughter Rachel had married and needed some furniture. We had always had old furniture in our home and although you don't realise it, your children take that taste with them and they wanted to acquire some furniture similar to ours. So I was asked to go to the auction with them and possibly give them some advice.

We decided to go to an auction, which turned out to be quite a good one with some nice pieces there. We established ourselves and started to look round. My son-in-law saw a nice set of chairs which he liked but they were a late entry and had not been catalogued so I advised him to steer clear of them. I thought they were not a very high quality chair and the source was unknown, so I thought this was good advice.

We went round looking at other things and marked down a nice set of drawers, some cupboards and my son-in-law had for some reason earmarked a stag's head with huge antlers. I was a bit taken aback and said he would have to find a large place to put it, but he wanted it for a hat stand! I thought to myself: well, everyone chooses what they want to do in life.

A lot of the furniture had come from a good home and you could see someone had loved it and it had been a possession of pride and I always feel if you buy something like that, you take on a responsibility to look after it in turn, but those are just my thoughts. Certainly with the pieces I've bought at auction over my life, I've followed that thought on.

The auction was beginning to get underway and in due course the set of ten chairs came up. They started at very little money, maybe two hundred pounds, but they soon went up to six hundred and my son-in-law bought at that price.

I thought to myself, 'Well, the deed is done.'

We went to collect the chairs and I noted the mark on the bottom of the chairs was 'Gillows of Lancaster' and they were the makers to the Queen, with a Royal Warrant. If I had ever been wrong in advising my son-in-law I was one hundred percent wrong this time! He had bought them for six hundred pounds and my rough estimate was that they were worth six thousand pounds. I could not believe it: the only explanation I could come up with for the lack of interest in them was that because they were a late entry, everyone had missed their value – but my son-in-law had won that one. They also bought the other pieces they were interested in, including the stag's head!

Next came the task of loading them onto the trailer. We managed to put the chairs, the chest of drawers and the other small pieces on board and that left the stag's head. The only place that would fit was in the cab wedged between us with the antlers sticking out of the windows. It was a huge stag's head and it took up most of the room in the cab with me flattened into the corner so as to give Andy more room to drive.

On reflection it was a very foolish thing to do, but then we are all guilty of foolishness from time to time. About three miles down the road we heard a large bang and we had lost the wardrobe off the back. Fortunately nothing was following us. I jumped out and started to pick up doors, shelves and fittings and whilst I was doing this, a police car arrived. They wanted to know what had happened and I explained that one of the ropes had slipped. He told me we had an unsecured load, and that they had been in a lay-by about half a mile back and had been watching us as we drove by but they could not figure who the third person in the cab was.

I could hardly keep my face straight as he walked round to the cab to investigate. Fortunately he was a policeman with a sense of humour and he burst out laughing. He said on this occasion he would give us a caution for what he termed our 'Laurel and Hardy' style. We secured the load extra carefully. I think the stag's head saved us from a police ticket, so maybe I won't mock my son-in-law's taste in the future!

A call came from a television company who wanted to know how we smoked our bacon. I told the young lady on the phone we were traditional smokers and we smoked the bacon in a variety of ways with oak, apple and beech wood and with different herbs. She said they were doing a programme on smoked bacon which had had the smoke painted on, and I told her that method had been around for a long while but we did not participate in that method, as I did not think that that was the correct way for a traditional product: it was a bit like producing under false pretences.

The young lady agreed with me and asked if they could come and do a programme with us and she said they would be bringing a well-known Irish chef with them to cook some of the bacon, if it was convenient to use our kitchen. I said she could come and look at the kitchen but first of all I would have to get permission from the wife!

On the appointed day, the TV people arrived with the camera crew and back-up team. The producer was an amazing sight: she had on what looked like a long nightdress and it was, in my opinion, a nightdress; and a pair of high heeled shoes. She was a very attractive young lady but I could not understand why she wanted to look like she had just got out of bed!

Later I was told it was a fashionable look but I still found it very amusing. The producer wanted to understand the smoking process, so I explained how smoking was originally done to preserve the meat or fish for the winter. I said that some of the old fishing communities around Britain still have local caves with coatings of tar on them, and for a long time people could not understand how this coating got there, but it was simply because the fishermen years ago smoked the fish in the caves, a ready-made smoke house. They used to light the fire in the cave, put the fish on racks and smoke it. I told her smoking food goes back to the Greeks, Egyptians and even further, back to the Persians.

She asked what food items could be smoked and I told her you could smoke most things: chicken, salt, cheese, bacon, garlic, fish, the list is endless. But it's not recommended to smoke fish in the same smoke house as meat products, because you get a taint from the fish to the meat. It is better to have a separate kiln for fish.

Basically speaking, smoking is a simple art but it has to be done correctly. Our kiln was a traditional English brick kiln but in other countries they are made out of different materials. In Scandinavia, for example, especially Sweden, kilns are made out of wood, as they are in some parts of America too. Other countries put a soil roof on them; a flat roof covered with soil then grass on the top, which acts as insulation.

Smoking is like sailing a ship: if the wind is right, you set your sails and it's all down to experience. You put all your products in the smoke house, hanging them from racks and making sure there is a good space between them so that all the smoke can circulate nice and gently.

The big modern kiln is different: the smoke is blown round under pressure and this makes the bacon hard. I am not an admirer of this method as I think smoking should be done gently, and the bacon kept softer. I showed her the kiln we had built, which had the fire box on the outside.

Traditionally, first of all straw was put on the floor and damped down, then the sawdust was put on the top; barley straw was always used as they used to say it gave the smoke a better taste. In earlier times, juniper and other herbs were added to the sawdust to give it a better aroma and the smoking was a profession: it was someone's whole job to be a professional smoker.

Today businesses buy an electric box and smoke the bacon over-night or they paint the smoke flavour on. How they do this is to buy a bottle of smoke-flavour from the butcher's supplier, lay the bacon on a table, paint the first coat on, let that dry and put a second coat on. It does look like smoked bacon but if you cut into the product about an eighth of an inch, you will find the 'smoke' does not go through the layers: it is only a top layer. With a traditional smoke, you will find the flavour goes all through the product.

In my opinion, the correct way to smoke is the traditional way using English oak or beech or apple wood to give a proper aroma and taste. A slow-smoke from any of these woods not only gives it a lovely flavour but it gives it shelf life and it gives the bacon that special taste.

The producer was very interested in all this background to smoking and the chef then took over and the cameras started rolling. My dog Dilly had taken a fancy to the chef and followed him everywhere, generally making a nuisance of herself, but for Dilly it was love at first sight. It was mutual because the chef, Paul Rankin from Northern Ireland, also took a liking to her.

The first shoot was of the smoke house and how it was set out, with the large hams in the centre and the smaller middle-cuts and the streakys all round the edge, enabling an even smoke. Paul described the different smokes and the difference between the traditional way of smoking and the short-cut, painted way.

Next he decided he would cook some of our bacon. He took two frying pans and cooked it, and to be honest, he made a terrible mess in the kitchen with grease everywhere. They did a blind tasting and the experts all agreed there was a considerable difference between the traditional bacon, which gave a deep, satisfying taste and the bacon with the smoke painted on. Personally, in shops I think there should be a notice on the painted products to say the smoke has been applied with a brush.

The crew carried on filming and finally Paul Rankin looked down at Dilly and said to her, 'Right, you have followed me all day long: now is the time for you to have your five minutes of fame. You can follow me down the yard as I take the bacon from the smoke house.'

But Dilly, being female, decided she did not want to do that. She sat down and refused to budge! No amount of persuading would make her follow him, so she never did get her five minutes of fame.

The shoot was over and they told me they had enjoyed themselves and they had learnt a lot about the bacon. The producer dug into her bag and brought the fee out; but my wife stepped in very quickly and commandeered the fee, telling me later that it would go towards cleaning the kitchen up!

CHAPTER 11

A Taste of Italy
and all about British Bacon

A large company contacted me with a view to producing Parma ham. They wanted me to go to Italy to see how it was produced there. I was very reluctant to go but fortunately Ann's daughter lived in Italy and spoke fluent Italian, and Ann can also speak a little of the language, so I realised that would help. We decided to accept the challenge and we thought we would make the journey by road. We organised the route, imagining it would take a couple of days, travelling via Rheims, Lyon, Genoa and down to Florence. I had always wanted to visit Florence.

The journey was organised, the car was packed and we set off for Dover. We caught the boat to Calais and then set off down the French auto route, we filled the car at the French service stations, noting the price of the diesel was a lot cheaper, at least half price.

We stayed in a very nice hotel and from memory we had dinner, breakfast and a moderate room for about seventy pounds for the two of us. I thought that was very acceptable. We carried on down to Lyon and stayed the night, then went on to the Frejus tunnel through the Alps and into Italy. We were asked if we were returning, which we confirmed, and we were duly sold a return ticket. On meeting Ann's daughter, she told us the ticket was only valid for five days, so we had been had!

We arrived in Florence and stayed in a very ancient hotel. For my own interest, we went round all the markets. I could not believe the relaxed hygiene standards in the main meat market: there was a man there with a huge cleaver and a big cigar in his mouth, chopping on a wooden block.

I thought to myself, 'Yes: where are the EEC rules and regulations in Italy?'

At that time, wooden blocks had been banned by the Commission as being unhygienic and we were having to use plastic boards. A foolish decision as properly maintained wood is more hygienic than plastic. We wandered round the market looking at all the different types of produce, and came across a dairy stall where, believe it or not, there were large plastic dustbins with cream in them, for sale to the public!

I thought to myself, 'This is unbelievable!'

We stayed in Florence for a couple of days and it was very enjoyable. The culture is different, especially the drivers, and traffic lights mean very little! One lunch time we decided we would like a nice meal. We went into a restaurant which was very busy but we were left to be served until last: the waiter kept ignoring us as it was obvious we were tourists.

When eventually the meal arrived, it was excellent, so it was worth waiting for. When the bill arrived, we had been overcharged. When Ann called the waiter over and queried the bill in Italian, the waiter was greatly taken aback and the bill was swiftly corrected.

After we had the break in Florence, we made our way back up to Parma and met Ann's daughter and grandchildren.

We headed off for the Parma ham factory, and on arrival we went into the foyer. There stood a tremendously old slicing machine and I had never seen anything more beautiful in my life; it was a make I did not know but somebody had put love and care into it, maintained it and looked after it and I thought this would mean the products would also be right.

We had a letter of introduction so I gave it to the receptionist; no more to do, a gentleman appeared and we started the tour of the factory. We went into where the hams were processed. It was amazing. I looked at the salt and the process, and was introduced to the Master Salter. Ann's daughter translated for us and she told the Master Salter that I was a Master Curer, and we shook hands.

Through translation I asked if the salt was the Roman salt and he nodded. He also explained how the salting machine worked and how it had been noticed that, after working a long time in the salt, the operators developed arthritis, and I thought to myself, 'It's just not an English disease.'

We went into all the different departments, I saw the whole Parma ham process and then we went into where they made all the salamis.

It was very similar to the salami method I had seen made before, but slightly different. The basic principal was the same, but they coated them differently to the ones I had seen: the Italians coated their salami with rice flour, which had been made into a paste and spread on the outside of the salami. This particular salami was made with rough-cut pork, and the salami man and I talked about the different types of salami: the German one, which is a fine, the Danish one which is different again and the Hungarian one which is a mixture of pork and beef.

Salami is made differently in every country in the world and in Italy, they must have ten or more different types, often named after the cities or areas in Italy it originated from. The Parma salami specialist was an interesting man to talk to and we enjoyed swapping one tale for another. There is no taking anything away from them, they made an excellent product and had some excellent ideas.

As we toured the factory it was explained to us about the micro climate in Parma and that is why Parma ham turns out as it does: it is a unique area and it produces a unique product. The area of Parma is surrounded by mountains; the air is warmed and remains trapped in this area, producing the perfect climatic conditions to make Parma ham (prosciutto crudo) the most famous in the world. You can't replicate that climate.

It takes up to two years to produce a Parma ham. The hams all hang in specially designed rooms to catch the warm air and aid in the curing process; it is the most amazing sight to see hundreds and hundreds of hams suspended, drying in the aerated rooms.

We carried on round the factory, and finally came to the despatch area where I was amazed to see how many places in the world the Parma ham was sent to. It was an excellent product, an enterprise run with a lot of pride in the product. That is what you need in business: it should not always be about how much money you can stack up. It's about producing good food for good people, and a lot of the British food industries have lost that.

Other countries produce their own hams. The French produce a Bayonne ham which is nutty and sweet and is cured for a long period of time; the Germans cure their Westphalia ham by dry salting it and smoking it in beech and juniper berries; different countries all have slightly different methods and that is what makes the food industry so interesting.

We continued round the factory. I had learnt a lot over the years and I thought my team and I had produced some good ham and bacon, but true wisdom is when you realise you have still have a lot to learn. I thought about how differently the Italians process the pig than we do and there was no doubt in some of the areas, they make a better job of it than us and I was grateful to see another point of view.

In England today I read about all these food award ceremonies and some of these people who are new to our industry think they are the only people who have discovered good food, but they are wrong, we have always had good food and the talent is finding where it is still produced. Many the people produce good food but never shout it from the house tops and tell everyone how wonderful they are.

We finished our tour and arrived back in the reception area where it had been arranged for us to taste some of the salami. They brought out some lovely bread and a bottle of wine. The salami was absolutely delicious and it finished our tour of the Parma ham factory on a very nice note. It had been an enjoyable and instructive couple of hours; needless to say we visited the factory shop and bought quite a few supplies.

Next on our tour was the parmesan cheese factory, but on arrival there we were first asked if we would like to look at the pigs. The lady guide told us they were fed on the whey from the cheese-making and any surplus cheese and corn; they were certainly healthy-looking pigs and they had the look of the Neopolitan breed, with longer snouts than our English breeds.The flavour of the Neopolitan meat is excellent.

These Neopolitan pigs were huge: I would call them over-weights. The pigs in England are about two hundred and ten pounds live weight but these pigs must have been twice that weight.

We had a look round the cheese factory, and it was very interesting watching the cheese-making process. We ended up in the factory shop and bought enough parmesan cheese (parmeggiano) to open a shop! Altogether it had been a very interesting day, but the last visit on our tour was the winery!

We set off along the winding roads and arrived at the winery. They made us very welcome and we set off to look at the grape vines and were given some information on the grape varieties they used. Next we toured the factory where we looked at the process for extracting the juice from the grapes. In the enormous vats, the juice was fermented. This winery produced an expensive wine which was only sold to high class restaurants and in their own winery (cantina) shop.

After the tour we were shown into a large room with an enormous table and were invited to sit down. They produced the biggest platter of Parma ham I have ever seen, with lovely Italian bread, a couple of bottles of wine and water for the grandchildren. The ham was out of this world and we really enjoyed it, especially the children as you would think we had not fed them for days! We placed our order for the wine and while we were eating, it was loaded into the car for us. The car now looked like a travelling shop with not much space left for the travellers! I thought to myself: what a truly hospitable people the Italians are: I look on that day as a glorious experience of our travels in Italy.

We set off for home. I sat in the car contemplating our industry, with all its ups and downs, and I thought to myself as we rolled through the Italian countryside: what a wonderful career curing is, for people of all walks of life who want to produce good food and I felt that, if I was starting again, I would like to try to produce a version of the Parma ham.

We carried on to the Italian/French border and produced the return ticket for the Frejus tunnel only to be told it was only valid for forty eight hours. How could you go to Italy for only forty eight hours? I thought it was a con, but we had to pay all over again. The moral of that story is: if you go to Italy (via the tunnel), do not buy the return ticket! But the trip was an excellent experience. I could not have done the journey without Ann, as she had done most of the driving: she is a very competent lady.

We arrived home and everything was in order. I knew Ann had a special birthday coming up and I thought I should mark the occasion with a special present. Flowers and diamonds are a girl's best friend, and I thought in this instance flowers were not enough but diamonds might be. I decided we would look out for a good jewellery sale, and when the next one came up, we decided to go to the auction. I always enjoy looking at fine jewellery, especially at the craftsmanship.

The sale included rings, necklaces and silverware. I had a look at the items for sale and there was a silver locket which I opened up and in it was a lock of hair and the inscription '1914: something to remember me by'. I thought this locket should not be in a sale as it was so personal, and I did not like that. I put it down gently and walked away. In the sale room, there is normally someone in charge of the jewellery and you have to ask for the pieces to be given to you from a cabinet.

I particularly liked a solitaire and there was also a cluster of diamonds. We asked the lady to show us the rings and Ann tried them on. I personally liked the solitaire best but when I asked Ann, she said she liked the cluster ring.

The auction began and to this day I remember the lot number: it was 135. When the auctioneer came to Lot 120, I told Ann I needed to visit the toilet. She looked worried as our number was very near, but I told her not to worry.

I was in the toilet when I heard this little voice shout, 'Are you going to be long?'

She had no intention of me missing the bid!

I returned to my seat and Lot 134, the solitaire came up and that was sold, then he came to the cluster ring. He asked for bids, someone made a bid, then I made a bid, then the person from the other side of the room bid and I knew there was going to be a battle for this ring. We carried on bidding and I knew I was going to stick with it. I put in the last bid, the auctioneer banged his gavel and it was ours.

We went and collected the ring and Ann put it on straightaway and it did look beautiful. The choice of the cluster had been the better decision and I knew she had a beautiful ring and a token of love and happiness.

One day the BBC rung up and on the phone was Derek Cooper from the *Good Food Programme* who said he wanted to record a programme for posterity and he believed I was the last fully apprenticed Master Curer in England and I told him that was correct. He said he understood I produced all the regional bacons and I confirmed that even if we didn't produce them all at the same time I have produced them over the years. He told me he wanted to send his presenter down – a lady called Sheila Dillon – also a man called Peter Gott, a Cumbrian farmer who cured bacon from wild boar. I said I already knew of him and I would be very pleased for him to come along. Derek Cooper said it was important to record on television the traditional ways of curing bacon because as time goes by, this art would be lost and he felt I produced the 'Rolls Royce of bacon' and in his opinion I was, he said, 'one of England's National Treasures!'

I told him, 'Well, that is very kind of you but there are other people in the industry who also produce good bacon and I don't wish to appear arrogant. You can say it, but I can't.'

Sheila and Peter duly arrived and I encouraged them to look at the past and see how it had been done.

Originally, everybody in England with any land had a pig and that was their supply of meat for the winter. It had to be cured right as they only had one shot and if they mis-cured the pig, they would go without meat for the winter months. There would have been no refrigeration in those days so everything relied on the curing and in the majority of cases, there would have been a Master Curer journeyman to do this job. It was their job to make sure the pig they cured lasted.

This is a Master Curer who travelled from place to place, curing people's pigs for them, and in most cases he would have stayed with the family for a couple of days. He would arrive at the farm and they would kill the pig on the first day and dress it, then on his second visit they would cure it.

Tales are passed down, and often I have been told these Master Curer journeymen were colourful people and many left more than just bacon behind!

They used old English salt and saltpetre. The pig was cured in an old wooden trough covered with salt and it was cured for about a month. The meat was then washed off, patted dry and then hung in a barn to air-dry and that is when the flavour matured. It would take about six weeks to cure a pig from slaughter to end product and if a different taste was required, the legs were taken off and hung up the chimney to smoke. Basically that is how English people for hundreds of years had their supplies for the winter.

When the Master Curer journeyman had salted all the pig down, the ladies of the farm would process all the other by-products: the fat was salted and turned into lard and that would last a long, long time. The chitterlings (intestines) were salted, the head was made into brawn, the trotters and the tail was salted, the blood was turned into black pudding; the only thing left was the squeak! So the ladies on the farm produced all the by-products and the hams. In the majority of cases they produced the butter and cheeses along with the hams.

When the hams were fully matured, some of them were sent by coach to the London Coffee houses or were taken to the local butter market for sale. The ladies produced all these things, which were a considerable asset to the income of the farm. Some of this has been forgotten over the years and some of the recipes I have acquired have been found in old Bibles or in the back of old grand-father clocks, and we have the ladies to thank for preserving all these recipes.

Still focusing on the past, I explained how the Wiltshire bacon was sent to London as 'green bacon', in other words, unsmoked bacon. In London there was a huge trade in smoking bacon, using in particular juniper among the different herbs in the London smoke. There were shops in London which only sold bacon and ham, and they would have all the different regional hams. There also used to be shops selling nothing but sausages and it has all been forgotten.

Basically, a London ham shop could have York hams, Suffolk hams, Braden hams, not forgetting Ireland because there were the

Belfast hams which were a business within a business.

The taste of the bacon all depended on the breed and the food the pig had been fed on and in many cases those pigs were not kept in sties except in the deep winter and the rest of the year they were turned out into the vegetable garden to finish off the vegetables that had run to seed and then into the orchards to eat the fallen apples. The pigs turned the land over and cleared all the ground ready for the next season's planting and in this way the pig had been a good friend to the people of England. I have looked after pigs for sixty years and in curing them I can count on one hand the times we have found a pig not suitable for human consumption. In my view, the pig is one of the cleanest animals you can ever process.

After this little peep into the past, the producers asked me what kind of smoke I did. We mainly smoked four varieties. First we smoked the traditional oak smoke but we only used English oak, as Welsh oak smoke is harsher and so is the Spanish oak. We made sure we used the mild, English oak. If it wasn't oak, we also used apple wood. We used to buy up the wood from the old apple orchards and matured it until it was well dried. I also liked to mix an oak and a beech together and add some juniper berries to the smoke. Maple was also used to give the bacon a sweet taste. I also showed them my traditional smoke oven.

Of the regional bacons I produced, we looked at the Staffordshire Black, which was from the north of the county where they had the mining industry, the steel industry and where they made all the china. All of this was hard manual work and the Staffordshire workers needed a good quality bacon to give them energy. The Staffordshire Black was cured with dark treacle and sugar and its sweetness gave them energy.

In the pottery industry, once the huge bottle kilns were fired, the firers were not allowed to leave it, so they used to have to sleep round the kiln until it was ready. The kiln firers used to have a boy to fetch the beer and a boy to cook the meals, and the majority of the meals were cooked in the bottle oven.

They cooked the bacon on a shovel where it cooked very quickly so you needed a bacon cut that could take the heat. They would also add eggs and cheese and served it all up on the Staffordshire oat cakes – that is a traditional Staffordshire breakfast. They used to put the bacon, cheese and egg on the oat cake and roll it all up and eat it like that.

We looked at the other regional bacons: the traditional bacon, we cured with the very large grain salt, Maldon salt, and it was heavily salted, cured for about three weeks and matured for three months.

The Devonshire bacon was cured with a very light sugar, salt-petre and cider using mainly local materials. The Cheshire bacon was very sweet, as nearby Liverpool was the port for importing sugar, so that commodity was readily available. The York bacon was very gentle with not a lot of sugar and a fine grain salt. There are two types of Wiltshire bacon; one is the very famous bacon which was produced originally in Calne in the 1700s using a dry salted mixture, and was given a Royal Warrant. As time progressed, another Wiltshire cure was developed, using a wet brine which speeded up the process of curing.

Then there is the Shropshire bacon, which had a lot of different spices in the cure; these spices came from Bristol because there was once a lot of trade between Shrewsbury and Bristol, via the river Severn. The wool went down the Severn and the spices and the sugar came up the Severn from Bristol, and that is the reason the Shropshire bacon is a spicy bacon.

The Midland bacon was the Tamworth bacon, again a spicy one and that was dry cured using a heavy salt. The Welsh also had their own bacon and there were two types: the Welsh smoked their bacon using peat which gave the bacon a very distinctive flavour, they also liked their bacon to be very fat. I produced a fat bacon that was mainly fat with only a strip of lean and many Welsh people came into the shop specially to buy this.

Sheila Dillon and Peter Gott were very interested by all these different, regional bacons and the differences between them. Peter then asked me if there was a Scottish bacon and I told him there was: the Ayrshire roll. This is a unique bacon to produce and it is cut very thin. It is produced by the dry salting method, in Ireland, and shipped into Scotland. I had the pleasure one day of meeting a man who cured the Ayrshire bacon and he was kind enough to show me how it was done. It was an excellent bacon: very, very lean and cut very thinly, with the rind taken off.

I explained how each area also had their own breed of pig. For example, the Tamworth pig, traditionally favoured in the Midland area, is a long-snouted ginger pig whose meat is excellent for curing. The Tamworth is the oldest pure English breed, descended from the old English forest pig. There was also the Cornish pig which is similar to the large black; the Lancaster pig, and which later became the Large White, a very successful pig which now thrives all over the world. There is also the Shropshire pig, which produces particularly good hams; the Staffordshire pig which is a very coarse pig and the meat has a strong flavour; and the Cheshire pig which is white. The Gloucester Old Spot is an orchard pig and the Wessex and Sussex were two breeds of black and white pigs: these today have been cross-bred and are now called Saddlebacks which were very popular in the 1940s. These are just a few of the many breeds in England. The Welsh also had their own breed, very similar to the Landrace pig. It originally came from Sweden and is an ideal pig for processing.

With the Victorian Industrial Revolution, people moved into the towns and the bacon industry took off and we started to have factories solely producing bacon including the sweet cure bacon in a brine, which had previously been a farmer's kitchen recipe. Today curing is a huge industry and great advances have been made. Without arrogance and with pride, I can honestly say we produce more varieties of bacon in this country than anywhere else in the world.

Cured mutton was also discussed. It was quite a delicacy, cured

146

using the dry salted method and then smoked; predominately produced in Wales and in parts of Cumbria and Scotland and the mature sheep were used.

Mutton is still being cured and smoked today by specialist producers. Once again the Welsh smoked their lamb with peat but in Cumbria wood chips were used. It was quite a delicacy.

It's important to keep some of the old food recipes alive and I am always pleased when I can pass a little bit of knowledge on, because basically speaking, we can all learn from each other.

In making our bacon I bought all the ingredients individually. It would be easy to buy a ready-mixed product by the sackful but that would make my bacon and sausages no different from anyone else's product and I think the whole point is to produce food with a distinctive flavour, not all the same, and that is the thing I try to do. I like to look into the past and bring a little bit of that into the future and I hope that in my small way I have succeeded in doing that

Unfortunately I have no one following me in my business because all the children have made their own lives now, so I was the last of the line, which Derek thought was a great pity, but I told him I was hoping to produce a recipe book one day, to pass on. There will always be room in the food industry for more good food producers and everyone can adapt and add their own personal touches to their recipes. I would like to unlock many doors and hopefully pass on the joy to other curers that I have experienced through my work.

Meanwhile, the programme went out and a few days later the BBC rang and told me they had had more interest in the curing programme than any other in the series and that there was massive interest in traditional English food.

147

CHAPTER 12

Problem-solving, Ramblers
& some Bad Buys

A firm contacted me from the south of England who supplied the top end of the market with very specialised foods for high quality restaurants and hotels. He explained they were having problems with the preparation of some of his pork products and suckling pigs and they wanted me to come down for a day and give them some advice, so I agreed to visit them.

I set off into the Home Counties, into the back of beyond, but eventually I found it, at the end of a winding lane; one of the most beautiful Georgian houses I have ever seen. It was surrounded by conifer trees and oaks and whoever had designed the surrounding gardens had a rare talent.

I drove into the cobbled courtyard, a man came out and we introduced ourselves, and he invited me into the house.

I stopped at the door and wiped my feet.

He said to me, 'You're early,' and I replied, 'Yes I am: that is the politeness of princes,' and he laughed.

'I know we are going to get on, as you wiped your feet!' he said, and I thought to myself he was a bit of an eccentric.

The house was huge and as you walked into the main entrance there were massive pillars on either side and a glass roof in the entrance and as you walked further in you came to the stairs, which were wide enough to drive a bus up.

He took me into a room he called a salon and we sat down and he told me a little bit about his estate. He supplied the top end of the market with specialist foods and he was having problems with the suckling pigs, because they were breaking up as they cooked. He also made haggis but was running into problems with those. Apparently, they sold a lot of pig's feet to the high class restaurants and I was a bit taken aback by that but he assured me they were classed in those circles as a delicacy.

He asked me if I wanted to look round. It was a huge enterprise and very well run. In one of the buildings there was a lovely factory. He showed me the cooked suckling pigs which were just cooling off. He asked me what I thought about them and I told him I would have to think about it.

He then took me to where the pigs' feet were produced and the problem they had with them breaking up. He also said many of the haggis kept bursting out of their casings. I told him if he left me with the factory manager for a while, and let me have a look round, I would do what I could.

He agreed to that, so I set off with the manager and had a close look at all the production lines. I looked at the pig's feet production and could instantly see where the problem lay. Next on my list was the haggis; I thought I could see the problem there. About an hour had passed and I thought I had solved the problems, so I told the manager I would walk back to the house and he said he would walk with me. As we walked back, he told me a little bit about the house and their production line and I thought it was a very interesting place.

We met the owner again in his salon and he asked me if I had sorted the problems out. I told him I thought I had. I started with the suckling pigs. First of all, you wash them out with a very strong salt water solution to make sure they are clean, dry them off and then paint them with brandy. The next thing to do is to rub garlic all over the skin.

Then you make the stuffing with a rice as well as your sage, onion and salt, and you line the suckling pig with it. Then you sew the suckling pig up with strong cotton twine, leaving an inch between each stitch. You then lie the suckling pig on the wire tray and paint it all over with melted butter and salt. Repeat this process half way through cooking to produce a better colour.

The next step is to have a small block of wood, either sycamore or beech, and you place this block in the pig's mouth, then you make a hood of greaseproof paper and place over the pig's head which prevents the ears and snout from browning too much. When the suckling pig is cooked, you remove it from the oven and only when it has cooled do you remove the block of wood from its mouth and replace it with an apple, to dress the pig. On reviewing their methods, I suggested they cooked the suckling pig on a wire tray in the baking tin and cooked it at a lower temperature. The reason for this is that there is a high percentage of water in a suckling pig and it was being boiled rather than roasted and was being over-cooked in the process.

As far as the pigs' feet were concerned, I told him they should not be boiled; they should instead be steamed. All you need is a step in your boiler, a piece of stainless steel, shaped like a step, with holes in it. You put the pigs' feet on this, above the water level, to steam. When they have been cooked, leave them to cool still in the boiler and they will come out in one piece.

He asked me about the haggis. I told him I thought the casings were stuffed too tightly, and I suggested he added some flour to bind the mixture, then leave the casing only three-quarters full so the stuffing had room to expand.

I said, 'You don't need me any more.'

He then told me he wanted to produce a traditional ham and asked if interested in assisting them? I told him there were a number of recipes for old English hams and it depended how far back in time he wanted to go. He asked me what I meant.

I said, 'Well, I have a recipe that is two thousand years old, which is a Roman recipe, but basically it is a Celtic recipe, written down by the Romans because the Celts were not very literate.'

He asked me if it would still be possible to produce the Roman recipe and I told him it would be, but it would be expensive to produce, as we would have to fetch in a lot of materials specially, to produce an exact copy. He appeared to become very excited at the idea. He also asked about making an old sausage. I told him there was a recipe for a very old sausage that was given to the soldiers in the Roman army as iron rations, and it kept for a long time and all the Roman soldiers had this sausage so they had something to eat when they had finished fighting.

It was the original salami sausage, which was smoked. On a dig on a Roman site, a large pot was found with holes in the bottom and they could not figure out what it was used for. But the pot was one of the earliest curing vessels, to enable the water to flow out of the pork. At first the archaeologists could not understand how it had been used; and on the same site they also found some of the Roman smoke ovens, because most towns had communal smoke ovens where they took all their own products to be smoked so they would last. Smoking was the first preserving method in the world and that's how it happened. He really thought it was interesting and asked me if I would make the Roman sausage for him and I said I would.

He asked me about my arrangements for the rest of the day and I told him that as it was getting on for five o'clock, I should be heading for home because it was a long drive back to Shropshire. He asked if I would consider stopping the night and giving him another day tomorrow? I said I would have to ring home first to let them know what was happening. I did so and there was no problem.

That was the first day over and he told me they dined at seven o'clock and he would send one of the staff meanwhile to show me to my room. I went to the car and fetched my case, climbed the huge staircase and went along a passageway to the biggest

151

bedroom I had ever seen. It could have made two large bedrooms; it was original in decoration even down to the jug and bowl in the bathroom. But I was pleased to see there was running water and I did not have to fetch my own! I had a sit down to mull over the day's events, freshened myself up and went down to dinner at the prescribed time.

There were about eight of us for dinner and he introduced me to everyone.

He said, 'Maynard is an unusual man with a gift for curing,' and I did feel very embarrassed.

The other guests were all from the same county. He insisted that I sat at the head of the table and I protested that that was not correct but he was most insistent, so I sat down at the head of the table and we started to dine. There were a lot of staff in the house, the table was huge and could easily have taken twenty people. The evening was very enjoyable and I retired to bed.

The next morning we had agreed on an early start and we started to map out how we would produce the Roman ham and the sausage. I told him we would need to send to Italy for the ingredients, including the salt and the wine, which would have been a red wine. That would take about a month to arrive and he was happy about that. We went through the different stages of the preparation and how the factory would be organised to cope with the extra work. He was a very intelligent man and by having these extra lines, it would give the business a boost, and it would be an unusual item on the menu.

We agreed that I would produce the Roman meat from my own home. Eventually we were ready to start to produce the Roman hams. As luck had it, we had some contacts in Italy. Ann was going on holiday there and came back with the correct Roman salt, the original honey, some red wine and the Italian olive oil. We used our own pork. We also obtained the authentic ingredients for the Roman sausages and so we went into production.

We decided to cure ten Roman hams. It took us a while to salt ten down and then to mature them. We put them in salt for five days, took them out and restacked them, and then left them for a further twenty days, fetched them out and washed them off and dried them, put olive oil on them and then after a few more days of drying out, smoked them and then put more olive oil and vinegar on them and then left them to mature. They looked and smelt wonderful and I was very proud of them.

For the Roman sausage, we used pork, coriander, wine, salt and pepper and a little bit of olive oil, we put them in large casings and sealed them up. Then we put them into a strong solution of brine for a week, then fetched them out of the brine for a week and let them dry. Finally, we smoked them for four days. When the sausages came out of the smokehouse, we rubbed them with olive oil and let them mature.

We tested both the ham and the sausages and I can honestly say they were excellent. We had managed to make them with little change over two thousand years later, as we had cured them as the Romans and the Celts had. And the basic technique hadn't changed to how we do it today, so it must say something about the quality of good food – it has stood the test of time.

We despatched the ten hams and the sausages and I rang the specialist supplier up and told him I had cured the first ten and the sausages and they were successful and I would send him all the recipes, the smoking times and methods and see what he thought. Within a couple of days he was on the phone and I will always remember his words: he described eating the ham and the sausage as like dining with God. I thought to myself: well, that's another job done. The man carried on producing them for years and they were very successful. I remember him with affection as he treated me with kindness and courtesy and it was one of my better days out.

Recipe for Suckling Pig

Wash suckling pig on the inside with a strong salt water solution to make sure it is properly cleaned, then dry off fully.
On the inside, paint with brandy. Rub garlic all over the meat.

Prepare stuffing: 2lbs long grain white rice, 3lbs onions, ½ lb rubbed sage, 8ozs fine salt, 2ozs white pepper,
Method: Boil rice, rinse with hot water and leave on tray to cool. Finely chop onions, fry in butter until golden brown, mix salt, pepper and sage, mix all ingredients together.
Line the suckling pig with the stuffing. Sew the pig up with cotton twine, leaving one inch between the stitches.
Melt 2ozs of butter, add a large pinch of salt and paint onto pig.
Take a small block of wood, beech or sycamore; place in pig's mouth.
Make a hood of greaseproof paper, oiling the inside. Place on pig's head when the meat is beginning to brown, so the snout and ears do not singe.
Cook at a medium heat until meat is cooked through. Once cooled, remove wood from mouth and replace with an apple.
To serve, remove from tray, split down backbone from head to tail and cut into six portions.

We had a letter one day from a company who produced a book called *The Best of British Men* and they asked if I would be interested in sending my details to be included in this book. I looked at the letter and I was intrigued but could not make up my mind if it was a joke or a scam so I decided to ring the number on the letter and ask for the editor. I did this and they asked me my name and I told them it was Maynard.

A gentleman came on the phone and said, 'Hello Maynard,' as if we had been bosom pals for years.

I said, 'Excuse me for ringing up, Sir, but why do you want me in the One Hundred Best Men?' and he replied that I had been recommended by a person of importance and he understood I still produced all the regional recipes for bacon and ham and also that I was the last fully apprenticed traditional bacon curer in England.

'You have kept the candle flickering when it was not the vogue thing to do, because you thought it was right,' he said.

I was a bit taken aback and I said, 'Well, I did it to pass something on because curing is part of people's heritage, and all I ever wanted to do was produce good food for good people.'

He said to me I had answered my own question and to fill in the form and return it to him and they would publish it. Many months later we received an invitation to the launch of the book in London. Ann and I went with great pride and met many famous men and women and I felt very surprised and honoured to be among their number.

I was walking across the yard one day when I saw a man with a large map in his hand, a little tiny man, and I asked him if I could be of assistance.

He asked, 'Where's the right of way?'

I was a bit taken aback.

'As far as I know we don't have a right of way across our yard,' I said.

'Oh yes,' he replied, 'It goes right down the drive and through the stables.'

'Since when?' I asked.

He told me it was on the definitive map. I had never heard of the definitive map and told him so.

He said, 'In 1949 the Labour government laid out the right of ways for walkers.'

I asked him if anyone who owned the land had been consulted.

'Oh, no,' he replied, 'It is our right to walk these paths any time of the day or night.'

I was so taken aback I was nearly rendered speechless.

'So there is a right of way straight through my property?'

He nodded and told me it was his right.

I asked him hadn't he got something better to do? I asked him where the path should be. He pointed to the fields and stables and told me the path would go across the fields and through the stables. I told him the field did not go anywhere and there was a river running through the field and there was no bridge over the river. He told me that would not be a problem as a bridge would have to be put in, as it was their right to cross.

He also told me the gate across the main entrance would have to be left open at all times. I tried to point out to him that it would compromise our security as we had been burgled a number of times.

'If anyone can come here, anytime, we won't be able to challenge them, as they will just say they are walkers.'

He said, 'That's your problem: it is our right.'

We discussed the need for a bridge across the river, and when I asked him who would pay for it, he told me: 'The taxpayer.'

'So the taxpayer is to pay for your hobby?' I said.

I told him I thought the world had gone mad. He told me the path would go through the stables and I told him the stables had been there for a hundred years but he was quite adamant that that was where the path was. I asked him how he was going to walk through the wall and he told me a hole would have to be knocked in the wall.

'Well,' I thought, 'The world really has gone mad. That's totally ludicrous. What kind of country do we live in, when you can start knocking holes in an old stable to walk through?'

He said, 'That's the law,' with a smug smile.

In due course we had a visit from a man from the council who was responsible for opening all the paths in Shropshire. I explained to him we had lived there for many years and that nobody had ever walked the supposed right of way before.

He told me a public right of way was on the definitive map and he explained exactly what the definitive map was: in 1949 all the rights of way had been laid out and I asked him who had decided where they would be: he told me it was the local parish councils.

I asked him what use the path was, as it did not go anywhere, and he told me that some of them went through people's kitchens, and in that case they had to divert the paths.

So I said to him, 'Money is spent on diverting paths that should not be there in the first place.'

I could not believe this lunacy was happening. He told me I could divert the path but it would cost me money.

In due course we had to have the path diverted which cost a lot of money and the only time it was ever walked was by the man with a small group of people to make a point. It was one of my greatest disappointments that someone could come along and walk through your home and you could not do anything about it and from that time on I did not feel like my home belonged to me anymore.

One day I was working in the factory and a young lady asked me to go and have a look at something in her car. I walked over to the car and she took out a lovely painting and she asked me if I liked it. I told her I thought it was a wonderful painting. She explained that she had one or two at home: her Father was the painter and she wanted to sell some of them.

I told her I would like to look at the other paintings and agreed to go and see them that afternoon. She thought I should know her Father had just come out of prison and that he had developed the art of painting whilst he was in prison.

I said, 'Well, that is the way of the world and it won't cloud my judgement on the painting.'

So in the afternoon I set off to where she lived. I knocked on the door and she invited me in: all the walls in the front room were covered in pictures and it really took your breath away, it was like being in a picture gallery. Her Father came in and we shook hands and he told me all the pictures were his own work and I asked if I could look round.

He asked me what kind of pictures I liked. I told him I did not understand much about art but if I liked it, I liked it. He explained he had always had a tendency for art and he told me he had been incarcerated at Her Majesty's pleasure and had had to spend a few years away from society but he thought his time was an opportunity to turn a disadvantage to an advantage and he put all his being into his pictures. I felt very honoured to be in his company.

I looked round the room at the many pictures and the one that caught my fancy was a Christmas scene. There was a coach and horse outside an old village inn, with children playing in the snow. The coach had six horses, there was a village green surrounded by houses; it was quite a large picture and it was lovely.

He told me I had an eye for art and that it had taken him many hours to do. Unfortunately it was the most expensive picture he had, but we agreed on the price. I thought it was expensive but you have to pay for talent and I decided I would put it over the fireplace. He asked me if I wanted to take it there and then and I said I would.

I arrived home and we decided to put the painting in a privileged position over the top of a fireplace so that everyone could see it and enjoy it. It was winter time and we used to light a roaring log fire in the evening and thaw ourselves out a little bit.

One evening, whilst sitting in my chair looking at my picture, I noticed the coach and horses had moved. They had slid down onto the village green. I rubbed my eyes to make sure I wasn't seeing things, but sure enough the coach and horses were on the village green! I stood up and went to have a look and sure enough – the coach and horses had slid down.

I took the picture off and laid it on the table and to my horror the whole picture was on the move: the children, the village inn, the houses. When I looked closely I could see the pictures had been cut out of Christmas cards. The fire had melted the glue and all we had was a little bit of paint left. It was one of those situations where you did not know whether to laugh or cry so we took the laughter route and laughed and laughed until we could not stand up. We had been well and truly had. We had a little bit of joy out of it, as we saw the funny side. He took a lot of money from us but the recompense was that it gave us a lot of laughter: we put it down to experience.

Many months later his daughter came into the shop and I asked her how she was and she said she was fine. I wondered if I could ask her a personal question and she said she did not mind, so I asked her what her Father had been in prison for and she replied, 'He was a con man and a forger.'

I thought to myself, 'That figures,' and had a quiet chuckle.

I was walking across the farmyard one day when a man arrived in a very nice motor and he asked me if I was the proprietor and I replied, 'I don't know whether I am sometimes,' and he laughed.

He asked if I could spare him a moment. He told me he was a dealer in Oriental rugs and wondered if I would look at some of his wares. I told him before he started that we weren't in the class of oriental rugs as we were poor at that moment and if we bought one I felt we would have to cover it over with newspaper as we could not walk on it! We both laughed and he asked if he could show me some of them and I agreed to look at them.

He brought them into the house and they were in a long tubular container and the rugs were lovely. He told me he only had the two left; they were called Baccarat rugs and the design on them was the Tree of Life. I thought they were beautiful. The salesman was in full force and told me everyone should own something of beauty and for once I agreed with a salesman. I did like the rugs, so I told the salesman if he gave me a good price I would buy them. We bartered a bit and then shook hands and I think we both felt we had struck a good deal.

Amongst his other interests he collected antique jewellery and he happened to have two very unusual gold bracelets which could be made into a necklace. He asked me if I would like to look at them. He showed me how it worked: you could have a double bracelet or you could turn it into a necklace. I explained I did not think I could afford both the rugs and the necklace but he was very insistent that I could. He told me to weigh the necklace on the scales to see how much gold was in it and I did and it weighed eight ounces. I thought it would make a lovely present for Ann, so in a moment of weakness I struck the deal.

On a suitable occasion I presented Ann with the bracelet and she was delighted and it looked lovely on. I had as much enjoyment as Ann had in receiving it and that is how it should be.

I had checked the hallmark and it looked a lovely bracelet but unfortunately the road to heaven rarely runs straight.

Ann was wearing the bracelet one day when she wasn't working and receiving many admiring comments, when she noticed some of the links looked discoloured and that there was a dirty mark on her wrist, so we decided to seek a second opinion. Ann took it to a quality jeweller in the town; he had a good look at it and he told her it was a fake.

She asked how it was a fake and the jeweller told her it had been coated with a very thin coat of gold and that if we had paid a lot of money for it, we had been robbed. It was base metal with a fine covering of gold, made in the Far East, even down to the fake hallmarks! She came home and told me we had been had and I said to her, all my intentions had been right and I would make it up to her at a later date.

We decided to look at the carpets. I thought to myself: we better have an opinion on them too. I knew one of my customers was a carpet dealer so the next time I saw him I asked him if he would give me his opinion and I would pay him for his time. He told me he did not want paying as he considered we were friends; he looked at the carpets and said, 'Yes, they are the genuine article. They are Baccarat carpets and they are perfectly good ones: they are machine-made, not handmade, but I presume you have paid a high price for them and they are not worth the amount you have paid for them.' I thanked him very much and thought to myself, 'You can't win them all and in this case I haven't won any!'

After that we made a rule never to buy at the door and I think I have been better off financially and better-tempered: that was a lesson I never forgot.

I was working in the shop one day when a man pulled up in the yard and parked his car, an unusual motor car. He was dressed in a light leather jacket, a big cream hat, a pair of high-heeled cowboy boots and a very tight pair of jeans. He came into the shop.

I said 'Good morning,' to him and he came over to the counter, he looked at me and said, 'What actually do you do here?'

I replied, 'We are traditional bacon and ham curers. We also have a traditional smoke oven.'

'Hmm,' he said, 'Does that mean it's any good?'

'Well, it's what people seem to want,' I replied. 'If they like it, they come back again.'

'Have you been doing this for a long time?'

'More time than I care to remember,' I said, and he laughed.

'I've just moved into the big house in the village and I've been told you produce good food.'

He then told me his name, puffing his chest out as he did so.

I said, 'And my name is Maynard Davies.'

He looked at me with a puzzled expression and he repeated his name again and then said, 'Don't you know who I am?'

I said to him I did not know who he was, and he said he was the lead singer with a rock group.

I looked at him. 'Oh! And do you actually earn a living at that?'

He said, 'What do you mean, actually earn a living at that? Haven't you ever heard of my band?'

I said 'I am afraid I haven't.'

'We've won golden awards,' he said.

I said, 'That must be very nice for you, but I am very sorry, I haven't heard of your band.'

'Have you got such thing as a television?'

'Yes, I have a television but I just haven't come across your particular aspect of the music industry.'

He looked like a man who had been dealt a severe blow.

'You're taking the micky.'

I said, 'No I am not, I have just never heard of you.'

'You must be the only man in England who's never heard of us.'

'I think there might be one or two more than just me!' I replied

He said, 'I am famous!'

I replied, 'That must be very nice for you, being famous,' but he seemed to lose his temper and he said, 'You country bumpkins make me ache. You sit behind the rhubarb sticks and when the leaves grow bigger, you can't see out.'

He stormed out, jumped into his car and drove up the drive like a madman. I could hear all this sniggering behind me and the girls, who had been in the rest room, came out laughing fit to burst.

'Did you really not know who he was?'

'No,' I replied, 'I had no idea.'

They assured me he was a very famous rock star but I had still never heard of him. I think I must have dented his pride a little, a country bumpkin not knowing how famous he was.

I would definitely have known him if he came back into the shop, but sadly we never saw him again.

CHAPTER 13

The End of an Era

It was time to start thinking about the busy Christmas cycle once again. It was also time for me to rethink my life. Everybody has a moment of truth and I knew it was time to set sail on a new course. I reflected on how lucky I had been to have chosen, quite by chance, a craft which I thoroughly enjoyed and it does you good sometimes to sit back and count your blessings. I had some ups and downs but mainly ups. Now seemed the right time to decide my next move and I had to think about retiring.

I had had 55 years in the industry. I had started from a very small beginning and had ended up with a reasonable reputation for producing good food. You can always tell if you produce good food: it's when you get no complaints; you may not get any praise but if you never get any complaints, you can guarantee you are doing it right.

So we decided to bring in the services of a firm specialised in selling companies. The man came along, we talked to him for a couple of hours, and he said we had a unique business, but a business that would need somebody to really love it. He could see it wasn't all for money, it was a lifestyle business. He said he would advertise the business and see what came along.

Within a couple of weeks we had a number of enquiries. There was one from a major company who said if they did buy it, they

164

were only interested in the bacon production. They had a huge bacon business and mine would compliment theirs, but I had the impression they would have closed my business down and taken all the equipment with them. I did not want that: I wanted someone to buy the business to carry it on.

We had an approach from another company who had a similar business to mine and would tie the two together. Another firm was interested which had been in the business a long time and we knew each other by reputation. They arrived early one morning, we had a coffee first, and explained they produced a different kind of product, mainly wet cure bacon. They had a look round the factory and said it was very nice, but again they said they would take the name and move the production.

Next the agent said he had a young couple from London who would like to come and have a look. Duly they arrived at the house: there were five of them; I was quite surprised they had come with his Mother and Father and their baby of about nine months old. We showed them around the house and they liked it. Next I took them down to the factory while Ann kept the baby in the house, as it was a working winter day and very cold down there.

They wanted to know about the business and the figures, but I told them I did not want to discuss figures first, but what we produced, until I was sure they had the feel for the curing side of the business. I was not sure at the time, but I did like the young couple.

They had a look round and showed great interest in the business and they thought it was what they were looking for. I told them it was a nice business and good staff and I explained you didn't have to go out for custom, the customers came to you. I pointed out it was hard work: every crust you had to well and truly earn, but with the right attitude towards learning, you will succeed.

They decided they would like to have another look round, so went back down to the factory and I explained about the machinery and what each piece of equipment did, I showed them the vacuum packer which takes all the air out of the bags the bacon is packed in, the slicing machines, the washing-up area, the saws we used to break the pigs down, the sausage machine.

I asked the young gentleman who was interested in the business what he did for a living and he told me he first went to London and started to paint and decorate and then went into kitchen fitting and then into the building trade. I told him the food industry was quite different. He asked me if I would teach him and I told him I would if he had an open mind, yes I would teach him. You need to take the curing information in a little bit at a time, but you must remember, it's a craft and you don't learn it just like that.

It's a skill but I said, 'If you can bake a loaf, you can do this.'

Over the years I have taught many people how to cure bacon. I always took a lot of pleasure in teaching other people; even more pleasure when they made a success of it. It's no use taking secrets to the grave. We should share knowledge and those who don't want to share knowledge don't know much about their craft.

As the family had come from London, Ann had put on a lovely meal for them, a real country welcome and I was very proud of her. They sat down and had a meal with us in the kitchen and I thought then that if they took the business on and worked together, they would make a success of it. I was coming to the end of my business life and I wanted someone like them to carry on where I left off.

The next thing I heard from the agent was the news that the young couple were keen to go ahead with the purchase of the business. They were keen for the sale to go through before Christmas but I knew from experience that this time of the year I would be far too busy to teach them in time for them to come in and cope with the Christmas rush, so I knew the best thing to do was for them to take over straight after Christmas. The business always shut for most of January, so that would give me time to teach them and for them to settle in well before the business opened again.

After Christmas, the first stage was to teach them how to make sausages.

I insisted husband and wife were both there as they would both need to know the skills in case one of them was ill. I told them how to prepare the sausage meat, putting it and all the correct seasonings in the bowl chopper, a large bowl with blades that go round and chop the meat to the right consistency.

Before I could stop him, the young man had lifted the lid off the bowl chopper and all the sausage meat flew everywhere, splattering the surfaces. He immediately put it down, I switched the machine off and I think I was a tad grumpy because I was tired and I told him he must never do that as the blades are going at a terrific speed and there could be a nasty accident.

The tuition went on for a few weeks and I taught them as much as I could: how to cure bacon, sausage making, how to slice, how to pack, the maintenance of the machines.

Everything went well, they had a good basic knowledge of everything and I had opened the door for them. I was tired: the year had been a hard one and I was ready for retirement. The date was arranged for the handover, the staff had been acquainted with the new owners and were all being kept on.

Ann meanwhile had been packing the house up as she had also retired from her job. It was a mammoth task as we had been in the house for a few years. One day Ann was packing china in the dining room when out of the corner of her eye, she thought she saw a cavalier standing there! He was wearing a pale blue outfit and had long golden curls, but when she turned her head he had gone! I thought to myself that the 'presence' in the house had come to see what the disturbance of our leaving was all about.

The moving day arrived, the removal men came and you don't realise how much stuff you accumulate over the years until you stack it up. Most of the furniture was going into storage as we were going to live in Ann's old house until we had found ourselves another home. The move had upset our dog Dilly as she could not understand why I no longer went down to the factory every day, and she in turn no longer went to sit outside the factory, almost as if she had retired as well! She had also lost her long term friend as Lassie had died a few months earlier.

The young couple had asked for the feral cats to be left and we could see the sense in that but when they asked Ann if they could have Dilly her answer to them was, 'You are more likely for Maynard to leave me behind than Dilly!' and that was true, as I had a special bond with Dilly. Out of all the dogs I have known, she was the most special and very precious.

The house was now completely empty and I walked round it for the last time, thinking of all the happy times and all the sad times there had been: but now was the time to start the next part of our lives.

Both of our cars were loaded to the maximum and Dilly was in the truck waiting for me. As I drove up the drive I looked back at the house and said a final farewell and thought to myself: I wish the young couple all good fortune and hope they make a success of it. They had a good product, a good name and they were young enough to make their way.

I took one last look. It was the end of an era but it was the right time to go as I knew I was slowing up. The most difficult thing in life is to be honest with yourself; and you have to be honest with yourself. We then drove through the gates, ready for the next adventure.

Postscript

I hope you have enjoyed reading this book. If you like the lifestyle I have had and you would like to cure bacon yourself, I am next going to write a full practical handbook on how to do it. It will unlock all the doors and if any of you would like to start a career in curing, the book will be of great assistance to you.

I would like to add that curing is a craft where women play a prominent part; in fact the best curers I have ever met are women, and if it had not been for women in bygone years, some of the recipes would have been lost because it was the women who kept the records in family bibles.

At this particular time there is a great need for people who want to produce good traditional food and I hope in my small way I can help some future curers to achieve that. So if you want to take a different road in your life, this might be the opportunity. Meanwhile, I hope you have enjoyed reading about my life in bacon in this and in my first book, *Maynard, Adventures of a Bacon Curer.*

Maynard

Also published by Merlin Unwin Books
7 Corve Street, Ludlow, Shropshire SY8 1DB, U.K.
01584 877456
www.merlinunwin.co.uk

Maynard – Adventures of a Bacon Curer
Maynard Davies

Always one to turn a challenge into an opportunity, Maynard took pleasure as a young lad in learning the skills of the old master curers of the Black Country and he shares with the reader the secrets of top quality bacon, learnt over a lifetime: the methods, recipes, smoking and curing. His passion for, as he puts it, 'good food for good people', is his motivation - made by experts, using the best ingredients, and cutting no corners.

Funny, wise and very human, Maynard's unsentimental tale will remind readers that old fashioned virtues of pride in one's profession, hard work, an open mind and a lot of optimism, go a long way.

£9.99

Prue's New Country Cooking
Prue Coats

'This new edition of Prue's book is hugely welcome in my house. Her recipes are practical, and dependable, soundly based on the very best seasonal British produce, and invariably delicious.'

Hugh Fearnley-Whittingstall

'I think Prue Coats is the greatest game cook today – and her recipes are marvellous too.' **Clarissa Dickson Wright**

£15.99